Undergraduate Lecture Notes in Physics

For further volumes:
http://www.springer.com/series/8917

Undergraduate Lecture Notes in Physics (ULNP) publishes authoritative texts covering topics throughout pure and applied physics. Each title in the series is suitable as a basis for undergraduate instruction, typically containing practice problems, worked examples, chapter summaries, and suggestions for further reading.

ULNP titles must provide at least one of the following:

- An exceptionally clear and concise treatment of a standard undergraduate subject.
- A solid undergraduate-level introduction to a graduate, advanced, or non-standard subject.
- A novel perspective or an unusual approach to teaching a subject.

ULNP especially encourages new, original, and idiosyncratic approaches to physics teaching at the undergraduate level.

The purpose of ULNP is to provide intriguing, absorbing books that will continue to be the reader's preferred reference throughout their academic career.

Matthew Benacquista

An Introduction to the Evolution of Single and Binary Stars

 Springer

Matthew Benacquista
Department of Physics and Astronomy
University of Texas at Brownsville
Brownsville, Texas, USA

ISSN 2192-4791   ISSN 2192-4805 (electronic)
ISBN 978-1-4419-9990-0   ISBN 978-1-4419-9991-7 (eBook)
DOI 10.1007/978-1-4419-9991-7
Springer New York Heidelberg Dordrecht London

Library of Congress Control Number: 2012951295

© Springer Science+Business Media New York 2013
This work is subject to copyright. All rights are reserved by the Publisher, whether the whole or part of the material is concerned, specifically the rights of translation, reprinting, reuse of illustrations, recitation, broadcasting, reproduction on microfilms or in any other physical way, and transmission or information storage and retrieval, electronic adaptation, computer software, or by similar or dissimilar methodology now known or hereafter developed. Exempted from this legal reservation are brief excerpts in connection with reviews or scholarly analysis or material supplied specifically for the purpose of being entered and executed on a computer system, for exclusive use by the purchaser of the work. Duplication of this publication or parts thereof is permitted only under the provisions of the Copyright Law of the Publisher's location, in its current version, and permission for use must always be obtained from Springer. Permissions for use may be obtained through RightsLink at the Copyright Clearance Center. Violations are liable to prosecution under the respective Copyright Law.
The use of general descriptive names, registered names, trademarks, service marks, etc. in this publication does not imply, even in the absence of a specific statement, that such names are exempt from the relevant protective laws and regulations and therefore free for general use.
While the advice and information in this book are believed to be true and accurate at the date of publication, neither the authors nor the editors nor the publisher can accept any legal responsibility for any errors or omissions that may be made. The publisher makes no warranty, express or implied, with respect to the material contained herein.

Printed on acid-free paper

Springer is part of Springer Science+Business Media (www.springer.com)

This work is dedicated to my parents Patricia and John Benacquista who instilled in me the belief that I could do anything if I put my mind to it. It is also dedicated to my wife Marcia Selsor who insisted I put my mind to work on astrophysics.

Preface

Although I have had a long interest in astrophysics, my training has been in gravitational physics. I returned to astrophysics by way of gravitational wave astronomy, whose primary sources are expected to be binary systems containing white dwarfs, neutron stars, or black holes. It became necessary for me to learn about the astrophysics of the birth and evolution of systems that give rise to these sources.

The best way to learn a new subject is to teach a course on it. This book has arisen from an introductory graduate course in stellar astrophysics that I taught in 2007 at the University of Texas at Brownsville. Many of my students were preparing to be gravitational wave physicists and had little or no prior coursework in astronomy or astrophysics. Thus, I needed a book that started from the very basics of astronomy, but very quickly focused on stellar evolution. Unfortunately, I found no text that met my needs, and so I began writing my own detailed notes for class. These have evolved into the present textbook.

The intended audience consists of upper-level undergraduates or first-year graduate students in physics. Although narrowly focused on stellar evolution, the subject is still broad enough that individual topics are not covered in depth. The book is organized into four sections—measuring stars, equations and processes, stellar models, and dynamical systems. I have tried to provide enough background in each area so that students will have the necessary vocabulary for more in-depth studies of any topic covered in the book.

I have benefitted from numerous conversations and discussions with colleagues at the Center for Gravitational Wave Astronomy, who have helped with developing heuristic arguments in support of the mathematical treatments of many topics in this book. I would also like to acknowledge the students in my courses from 2007 to 2012, who served as both test subjects and proofreaders for the text. Of course, any and all errors in the text are my own. The final editing and polishing of the text were done at the Aspen Center for Physics, whose peaceful environment and wonderful support staff have made this an enjoyable process.

Brownsville, TX, USA   Matthew Benacquista

Contents

Part I Measuring Stars

1 Classifying and Describing Stars 3
    1.1   Celestial Motions and Times 3
    1.2   Celestial Coordinates 5
    1.3   Precession and Epochs 7
    1.4   The Magnitude Scale 11
    Problems 12

2 Introduction to Binary Systems 13
    2.1   The Two-Body Problem 13
    2.2   The Orbital Shape 15
    2.3   Time-Dependent Orbits 18
    2.4   The Orbital Elements 21
    2.5   Spectroscopic Binaries 24
    Problems 28

3 Measuring Other Stellar Properties 29
    3.1   Distances and Parallax 29
    3.2   Temperature and Blackbody Spectrum 30
    3.3   Radii and Eclipsing Binaries 34
    3.4   Boltzmann and Saha Equations 38
    Problems 43

Part II Equations and Processes

4 Stellar Evolution Equations 47
    4.1   The Energy Equation 48
    4.2   Hydrodynamic Equation 50
    4.3   Composition Equations 51
    4.4   Virial Theorem 55

4.5	Total Energy	56
4.6	Timescales	58
Problems		61

5 Gas and Radiation Pressures ... 63
- 5.1 Gas Pressure ... 65
- 5.2 Radiation Pressure ... 67
- 5.3 Degeneracy Pressure ... 68
- 5.4 Internal Energy of Gas and Radiation ... 71
- 5.5 Adiabatic Exponent ... 73
- Problems ... 75

6 Radiative Transfer and Stellar Atmospheres ... 77
- 6.1 The Radiation Field ... 77
- 6.2 Radiative Transfer ... 80
- 6.3 Radiative Heat Flux ... 83
- 6.4 Model Atmospheres ... 85
- Problems ... 89

7 Nuclear Processes ... 91
- 7.1 Nuclear Fusion ... 91
- 7.2 Hydrogen Burning ... 95
  - 7.2.1 The p-p Chains ... 95
  - 7.2.2 The CNO Cycle ... 97
- 7.3 Burning Heavier Nuclei ... 99
- 7.4 Neutron Capture Processes ... 101
- Problems ... 102

Part III Stellar Models

8 Simple Stellar Models ... 107
- 8.1 Polytropes ... 107
- 8.2 Polytrope Solutions ... 113
- 8.3 The Eddington Standard Model ... 115
- 8.4 The Eddington Luminosity ... 116
- Problems ... 117

9 Stability ... 119
- 9.1 Thermal Stability ... 119
- 9.2 Thermal Instability ... 121
- 9.3 Thin-Shell Instability ... 122
- 9.4 Dynamical Instabilities ... 124
- 9.5 Convection ... 126
- 9.6 Mixing Length Theory ... 129
- Problems ... 131

10 Stellar Birth ... 133
- 10.1 The Jeans Criteria ... 133
- 10.2 Formation of a Protostar ... 137
- 10.3 Contraction to Main Sequence ... 142
- Problems ... 145

11 Main Sequence Structure ... 147
- 11.1 High-Mass Stars ... 147
- 11.2 Low-Mass Evolution ... 156
- 11.3 Late-Stage Evolution ... 158
- Problems ... 162

12 Compact Remnants ... 163
- 12.1 White Dwarfs ... 163
- 12.2 Neutron Stars ... 169
- 12.3 Pulsars ... 172
- 12.4 Black Holes ... 175
- Problems ... 178

Part IV Dynamical Systems

13 Binary Evolution ... 181
- 13.1 The Roche Model ... 181
- 13.2 Mass Transfer Stability ... 184
- 13.3 Unstable Mass Transfer and Mass Loss ... 188
- 13.4 Binary Evolution Example ... 191
- Problems ... 195

14 Star Cluster Dynamics ... 197
- 14.1 Cluster Timescales ... 197
- 14.2 Globular Cluster Structure ... 203
- 14.3 Globular Cluster Evolution ... 207
- Problems ... 211

15 Dynamical Evolution of Binaries ... 213
- 15.1 Dynamical Formation ... 213
- 15.2 Binary Interactions ... 216
- 15.3 N-Body Integration ... 218
- 15.4 Binary–Cluster Interactions ... 220
- Problems ... 221

A Useful Constants ... 223
- A.1 Physical Constants ... 223
- A.2 Astronomical Constants ... 224

B	**Atomic Properties of Selected Elements**	225
	B.1 Atomic Properties of Selected Elements	225
C	**Closest and Brightest Stars**	229
	C.1 Closest Stars	229
	C.2 Brightest Stars	230

Solutions 231

Index 259

Part I
Measuring Stars

Astronomy is a very old subject with a rich history of observing and classifying celestial objects. Over time, the distance to the stars has been understood to be much greater than previously thought, and our observing platform has been discovered to move. Furthermore, our measurements and measuring devices have become substantially more sophisticated. Nonetheless, astronomy retains much of the historical nomenclature and usage. Although all quantitative measurements can now be defined in terms of standard units, their use can still confuse newcomers to the field. Here we introduce the techniques and units for measuring the properties of stars.

Chapter 1
Classifying and Describing Stars

The first thing that is necessary to begin describing stars is to have a means by which we can locate stars in the sky so that they can be referred to by others. In other words, we need to develop a *celestial coordinate system*. Generally, when developing a coordinate system, we look for landmarks to which can anchor the coordinate system. We begin by describing some bulk motions of the more prominent objects in the sky. From there we describe the subtle motions of the stars and the coordinate system used to locate them. We conclude with a system for describing the brightness of stars.

1.1 Celestial Motions and Times

During the course of a single day, we notice that celestial objects move through the sky from east to west and return (nearly) to their starting point in 24 hours. The effect is an appearance that the sky is a rigid sphere with stars, planets, moons, and the sun glued to it. This sphere seems to rotate in a westerly direction, but in fact it is the earth that is actually rotating in an easterly direction. Continued observation of the sky reveals that certain objects are not fixed to the sky, but move with respect to the stars. In addition to changing shape, the moon appears to move from west to east with respect to the stars, so that after 24 hours, it is not back at its original location. Thus, the moon rises about 40 minutes later each day. This is due to the fact that the moon is orbiting the earth in an easterly direction with an orbital period of about 27 days. The sun also slowly appears to move from west to east due to the orbital motion of the earth. In this case, the timescale for the sun to return to its original position is one year. Therefore, every day the sun appears to move about $1°$ to the west relative to the stars. These three motions are used to define the basic time units of day, month, and year. However, subtleties arise due to the fact that all motions were initially measured from the non-inertial frame of the earth.

Fig. 1.1 The appearance of the sun and a star relative to the meridian for a sidereal day compared with a solar day

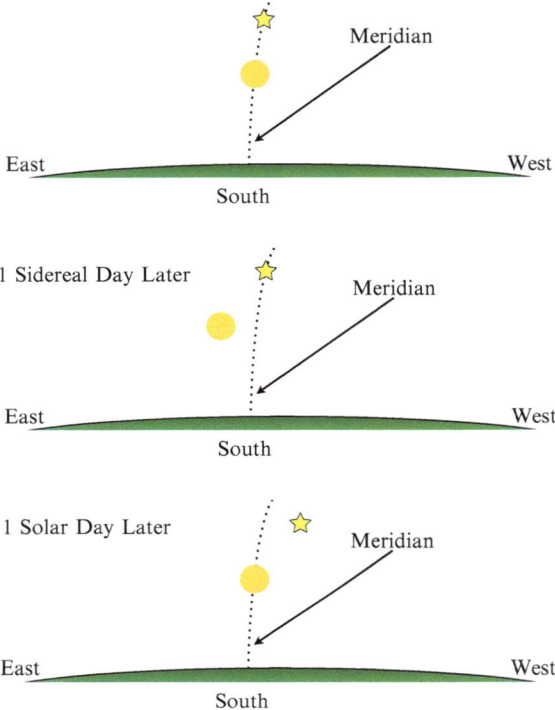

The standard definition of the *solar day* is the average amount of time required for the sun to go from crossing the *meridian* (the line dividing the sky into an eastern and western half) to returning to the meridian. Because the earth is orbiting the sun, the amount of time required for two successive meridian crossings of the sun is not the same as the time required for two successive meridian crossings of a star. The time required for two successive meridian crossings of a star is called a *sidereal day* and is about 4 minutes shorter than a solar day. Figure 1.1 shows the relative positions of a star and the sun on both a sidereal and a solar day. Because the earth's orbit about the sun is not perfectly circular, the actual time between successive meridian crossings of the sun varies throughout the year. When the earth is closer to the sun, the solar days are a little bit longer than 24 hours, and when the earth is farther from the sun, the solar days are a little bit shorter than 24 hours.

Although the standard definitions of the months have been manipulated for millennia due to a variety of political reasons, the underlying basis of the ~30 day month is the time between two successive full moons. This is called a *synodic month* and is about 29.5 days. Since a full moon requires that the earth, moon, and sun are aligned, a synodic month is longer than a *sidereal month*, which is the time it takes for the moon to return to the same location with respect to the stars. A sidereal month is about 27 days. The relative positions of the sun, moon, and earth at a sidereal and synodic month are shown in Fig. 1.2.

1.2 Celestial Coordinates

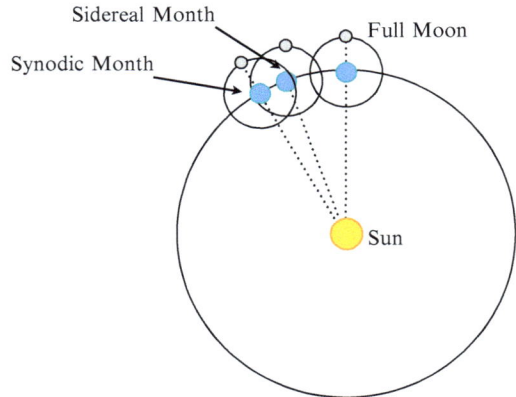

Fig. 1.2 The orientation of the earth, moon, and sun at full moon, a sidereal month later, and a synodic month later

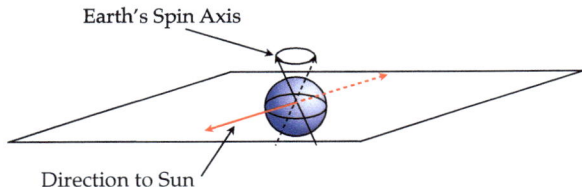

Fig. 1.3 The orientation of the vernal equinox at two times during the precession period of 27,500 years. The vernal equinox is a line joining the sun and the earth on the first day of spring. This line is also the intersection between the plane of the earth's equator and the plane of the earth's orbit. As the spin axis of the earth precesses, the direction of this line rotates in the plane of the earth's orbit. The *dashed line* shows the direction of the vernal equinox after 13,250 years

The definition of the *sidereal year* is the amount of time it takes for the sun to return to the same position with respect to the distant stars. A commonly used reference point is the position of the sun on the first day of spring. This point is known as the vernal equinox. Because the rotation axis of the earth is precessing slowly, the sidereal year is not quite the same as the amount of time between successive vernal equinoxes. It takes roughly 27,500 years for the vernal equinox to return to its initial position relative to the stars (Fig. 1.3). We will see in the next section that this *precession of the equinoxes* has an influence on the coordinates used to describe the positions of the stars.

1.2 Celestial Coordinates

Although there are a number of specialized coordinate systems for describing the locations of objects in the sky, the most commonly used system by astronomers

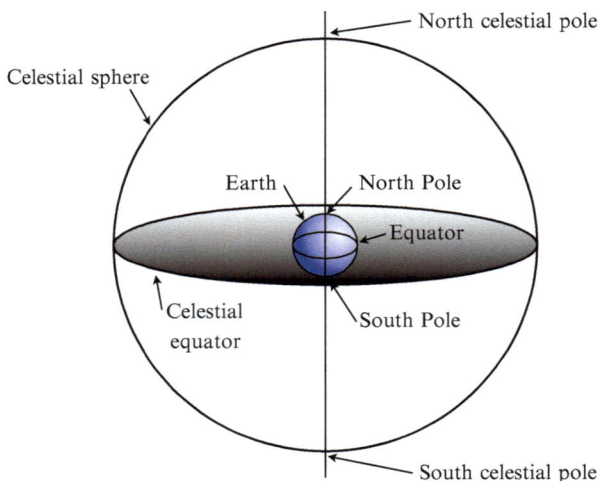

Fig. 1.4 The celestial sphere as an extension of the earth

are *equatorial coordinates* (also known as *celestial coordinates*). This system is essentially the angular components of spherical polar coordinates centered on the earth. As with any spherical polar coordinate system, we need to define two axes and a right-hand rule from which we measure the angles. The equivalent of the z-axis is the rotation axis of the earth and the equivalent of the x-axis is the location of the sun at the vernal equinox. A recurrent consequence of the fact that astronomy is an ancient science is that many conventions are derived from old concepts that made sense at the time. In this case, the actual angular coordinates that are used are based on the idea that the sky is modeled as a sphere surrounding the earth with all stars, planets, and moons stuck on the sphere. This sphere rotates around a fixed earth in a sidereal day. All the stars are fixed to the sphere, but the planets, moon, and sun move on this sphere. Equatorial coordinates are then the analogues of latitude and longitude on this sphere (Fig. 1.4).

The *declination*, δ, is the equivalent of latitude and ranges from 90° at the north celestial pole (NCP) to −90° at the south celestial pole (SCP). The celestial equator is defined by $\delta = 0°$. The declination is related to the polar coordinate θ (also known as the *co-latitude*) by $\delta = 90° - \theta$. Generally, δ is given in degrees, minutes of arc, and seconds of arc.

The *right ascension*, α (or RA), is the equivalent of longitude. Due to the daily rotation of the celestial sphere, the right ascension is measured in terms of the amount of time you would have to wait from the point that the vernal equinox crosses the meridian until the object of interest crosses the meridian. Thus, the units of RA are in hours (not degrees or radians). In order to standardize the units, one hour of right ascension is now defined to be exactly 15°. Generally, α is given in terms of hours, minutes, and seconds (of time — not arc). The right ascension is related to the polar coordinate ϕ (also known as the *azimuthal angle*) by $\alpha = \phi/15°$ (Fig. 1.5).

Fig. 1.5 Celestial coordinates

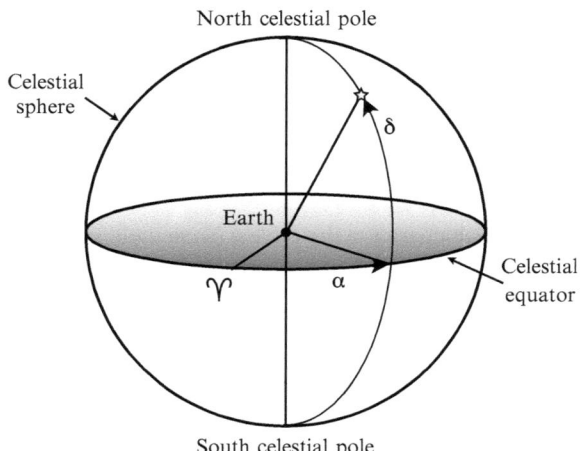

1.3 Precession and Epochs

Equatorial coordinates would be a wonderful and stable system if the earth's axis of rotation didn't move. Unfortunately, the earth's axis wobbles due to the gravitational interaction between the nonspherical earth and the sun and moon. The dominant period of this precession is 25,700 years. Due to the precession, the location of the north celestial pole moves along a circle in the sky that is $\sim 47°$ in diameter. Since the sun moves along the *ecliptic* (a great circle that is inclined by 23.5° with respect to the celestial equator) and the plane of the celestial equator is tied to the orientation of the north celestial pole, the position in the sky where the sun crosses the equator will change due to the precession. The vernal equinox moves at about 50.26″/year westward along the ecliptic. Additional earth–planet interactions add about 0.12″/year. Therefore, when we give the equatorial coordinates for an object, we must also give the time, or *epoch*, for which those coordinates are valid.

Over reasonably small timescales the change in celestial coordinates is linear (and most human timescales are small compared to the precession period). Thus, we specify the epoch or time for which the celestial coordinates would be accurate. The epoch is usually B1950 or J2000 which correspond to January 1, 1950 or 2000, respectively. If we are taking an observation on some other day that is N years after the epoch and want to point our telescope in the correct direction, we need to adjust the coordinates. For small N (generally less than about 50 years), the correction is linear in N:

$$\Delta \alpha = [m + n \sin \alpha \tan \delta] N, \tag{1.1}$$

$$\Delta \delta = [n \cos \alpha] N. \tag{1.2}$$

For conversion from J2000 coordinates, $m = 3.075$ s/year and $n = 1.336$ s/year $= 20.043″$/year.

For example, if we want to point our telescopes at Aldebaran (the giant orange star in the constellation Taurus) on the night of September 30, 2009, we first need to locate its J2000 coordinates. A very good web reference for finding the celestial coordinates of many stars (as well as other objects) is the *STScI Digitized Sky Survey*, located at

http://archive.stsci.edu/cgi-bin/dss_form.

Entering Aldebaran in the *Object Name* field returns

$$\alpha = 04^{\rm h}\, 33^{\rm m}\, 55.24^{\rm s}, \tag{1.3}$$

$$\delta = +16° \, 30' \, 33.5''. \tag{1.4}$$

After obtaining the J2000 coordinates, we need to determine the number of years that have passed:

$$N = (2009 - 2000) + 273/365.25 = 9.7474 \text{ year}. \tag{1.5}$$

Next, we convert α and δ to degrees (or radians) for use in the trig functions:

$$\alpha = (4 + 33/60 + 55.24/3600) \times 15 = 68.48017°, \tag{1.6}$$

$$\delta = 16 + 30/60 + 33.5/3600 = 16.50931°. \tag{1.7}$$

Finally, we put these numbers into the epoch correction Eqs. (1.1) and (1.2) to obtain

$$\Delta\alpha = 33.565^{\rm s}, \tag{1.8}$$

$$\Delta\delta = 71.665'' = 1' \, 11.665''. \tag{1.9}$$

Therefore, we should point our telescopes at

$$\alpha = 04^{\rm h}\, 34^{\rm m}\, 28.82^{\rm s}, \tag{1.10}$$

$$\delta = +16° \, 31' \, 43.6''. \tag{1.11}$$

Unfortunately, if we point our telescopes at this position, Aldebaran won't be there. In addition to the slow motion of the celestial coordinate system due to precession, the stars themselves move through space. Motion directly along our line of sight to the star does not result in any change in the celestial coordinates. This type of motion is called radial motion and can be detected through Doppler shifts in the stellar spectrum, as will be discussed later in Chap. 3. Motion perpendicular to the line of sight is called *proper motion*, and results in a change in the celestial coordinates of the star. Concentrating solely on the proper motion, we can define the two-dimensional projection of the three-dimensional velocity onto the celestial sphere in terms of a change in angular position with respect to time, $\mu = d\theta/dt$, where θ is the angle subtended by the star along a great circle. For a given time interval Δt, the star will move an angular distance $\Delta\theta = \mu \Delta t$. We can think of this

1.3 Precession and Epochs

Fig. 1.6 Spherical triangle

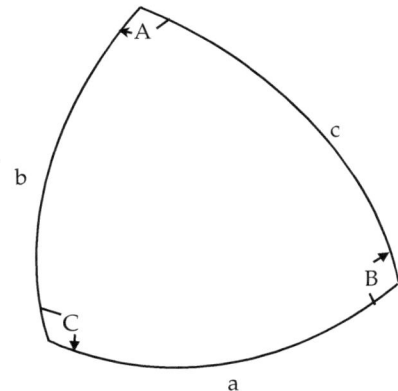

angular displacement as a vector lying on the surface of the celestial sphere and then break it up into components along α and δ directions. Since we are working with angular displacements on a spherical surface, we need to use spherical trigonometry and the spherical laws of sines and cosines. If we consider a triangle drawn on the surface of a sphere and define the angles (which are angles) and sides (which are *also* angles) as shown in Fig. 1.6, then the law of sines is

$$\frac{\sin a}{\sin A} = \frac{\sin b}{\sin B} = \frac{\sin c}{\sin C}, \quad (1.12)$$

the law of cosines for the sides is

$$\cos a = \cos b \cos c + \sin b \sin c \cos A, \quad (1.13)$$

and the law of cosines for the angles is

$$\cos A = -\cos B \cos C + \sin B \sin C \cos a. \quad (1.14)$$

Consider the triangles defined by the proper motion displacement and the position of the star on the celestial sphere, as shown in Fig. 1.7. The sides of triangle *PAB* are

$$\bar{PA} = 90° - \delta, \quad (1.15)$$
$$\bar{AB} = \Delta\theta, \quad (1.16)$$
$$\bar{BP} = 90° - (\delta + \Delta\delta). \quad (1.17)$$

Consequently, the law of sines gives

$$\frac{\sin(\Delta\alpha)}{\sin(\Delta\theta)} = \frac{\sin\phi}{\sin[90° - (\delta + \Delta\delta)]} \implies \sin(\Delta\alpha)\cos(\delta + \Delta\delta) = \sin(\Delta\theta)\sin\phi \quad (1.18)$$

Fig. 1.7 Proper motion along the celestial sphere

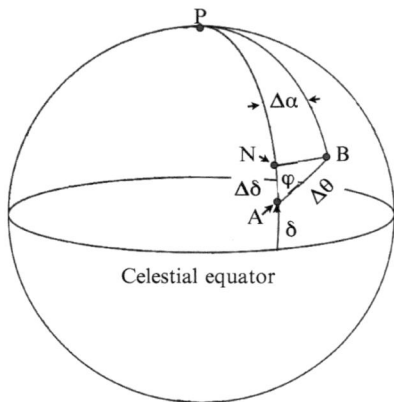

and the law of cosines for sides gives

$$\cos\left[90° - (\delta + \Delta\delta)\right] = \cos(90° - \delta)\cos(\Delta\theta) + \sin(90° - \delta)\sin(\Delta\theta)\cos\phi. \tag{1.19}$$

This looks pretty ugly until we realize that $\Delta\theta$ is always a small angle, and so $\Delta\alpha$ and $\Delta\delta$ are also small angles. Therefore we can use the small-angle approximations to obtain

$$\Delta\alpha = \Delta\theta \frac{\sin\phi}{\cos\delta}, \tag{1.20}$$

$$\Delta\delta = \Delta\theta \cos\phi, \tag{1.21}$$

$$\Delta\theta = \mu\Delta t. \tag{1.22}$$

Sometimes the proper motion is described in terms of μ and ϕ, and other times it is described using $\mu\sin\phi$ and $\mu\cos\phi$.

Problem 1.1: Use the SIMBAD database at

http://simbak.cfa.harvard.edu/simbad/

to determine the position of the star Capella on your birthday for this year. Give the new RA and dec after precessing and then the new values of RA and dec after including proper motion. Note that the proper motions are given as RA=$\mu\sin\phi$ and dec=$\mu\cos\phi$ in this database.

1.4 The Magnitude Scale

A quick look up at the sky with the naked eye reveals that some stars are brighter than others. This is such an obvious fact that a ranking scheme for measuring the brightness of stars was developed in classical times using the human eye as a photon detector. The scale used by Hipparchus ranked stars according to their brightness with the brightest stars of *first magnitude* ($m = 1$) and the dimmest stars that he could see were of sixth magnitude ($m = 6$). One odd thing about this ranking scheme is that brighter stars have lower magnitudes. Consequently, the brightest objects in the sky now have *negative* magnitudes. Another odd thing about the magnitude scale is that it is based on the human eye, which is *logarithmic*. With the advent of devices for measuring the brightness of stars, it became apparent that the magnitude scale defined by Hipparchus could be well approximated by a logarithmic scale where a difference of five magnitudes corresponded to a factor of 100 in brightness. This definition of the magnitude scale can be put on firmer footing by considering *radiant flux*.

The radiant flux, I, detected by an observer (using either eyeballs or more sophisticated instruments) is defined to be the amount of light energy deposited per unit area per unit time. Without defining the units that we shall use for I, we can still quantify differences in the magnitude scale in terms of ratios of fluxes. If we define the flux and magnitude for star 1 as I_1 and m_1 and for star 2 as I_2 and m_2, then

$$\frac{I_2}{I_1} = 100^{(m_1 - m_2)/5}, \tag{1.23}$$

and

$$m_1 - m_2 = -2.5 \log_{10}\left(\frac{I_1}{I_2}\right). \tag{1.24}$$

Note again that lower magnitudes correspond to brighter stars.

Another thing to consider about the magnitude scale (as well as the radiant flux) is that there is no consideration of the distance to the star in the definition. Consequently, two identical stars may have different magnitudes if they are at different distances. We define the *luminosity*, L, of an object to be the total energy emitted by it. If an object is radiating light uniformly in all directions, then we can relate the radiant flux to the luminosity and distance by the familiar inverse square law for light:

$$I = \frac{L}{4\pi d^2}. \tag{1.25}$$

If we can measure the distance to a star, then we can determine its luminosity from the radiant flux. We then define the *absolute magnitude*, M, to be the *apparent magnitude*, m, that the star would have if it were placed at standard distance away.

The standard distance that we choose is 10 parsecs where the *parsec* will be defined in Chap. 3. In standard units, the parsec is

$$1 \text{ pc} = 3.26 \text{ lyear} = 3.086 \times 10^{16} \text{ m}. \tag{1.26}$$

(See Appendix A for tables of standard astronomical units.) In terms of the known distance, d, to a star and its apparent magnitude, m, the absolute magnitude is

$$M = m - 5\log_{10}\left(\frac{d}{10 \text{ pc}}\right). \tag{1.27}$$

The *distance modulus*, D, is the difference between the apparent and absolute magnitudes, so

$$D = m - M = 5[\log_1(d) - 1]. \tag{1.28}$$

Frequently, the symbol μ is used for the distance modulus, but we have chosen D to avoid confusion with the proper motion. The distance modulus is useful when describing a cluster of stars that are all at roughly the same distance because one can simply add a constant distance modulus to each measured apparent magnitude in order to obtain the absolute magnitudes of the stars in the cluster.

Problem 1.2: Given that the luminosity of the sun is $L_\odot = 3.84 \times 10^{26}$ W and the absolute magnitude of the sun is $M = 4.74$, find the apparent magnitude of the sun. The distance to the sun from the earth is 1 AU $= 1.496 \times 10^{11}$ m.

Problem 1.3: The star Sirius is 2.64 pc away from the earth and it has an apparent magnitude of -1.44. What is its luminosity in units of the solar luminosity, $L_\odot$?

Problems

1.1. Use the SIMBAD database at http://simbak.cfa.harvard.edu/simbad/ to determine the position of the star Capella on your birthday for this year. Give the new RA and dec after precessing and then the new values of RA and dec after including proper motion. Note that the proper motions are given as RA $= \mu \sin\phi$ and dec $= \mu \cos\phi$ in this database.

1.2. Given that the luminosity of the sun is $L_\odot = 3.84 \times 10^{26}$ W and the absolute magnitude of the sun is $M = 4.74$, find the apparent magnitude of the sun. The distance to the sun from the earth is 1 AU $= 1.496 \times 10^{11}$ m.

1.3. The star Sirius is 2.64 pc away from the earth and it has an apparent magnitude of -1.44. What is its luminosity in units of the solar luminosity, $L_\odot$?

Chapter 2
Introduction to Binary Systems

In order to model stars, we must first have a knowledge of their physical properties. In this chapter, we describe how we know the stellar properties that stellar models are meant to replicate. Some of our data comes from observations of nearby single stars, but much of our information comes from binary stars. We will begin by describing the orbit of a binary and how these orbits are observed. We conclude this chapter with a discussion of how stellar masses are obtained from observations of the spectra of binary stars.

Binary systems are observed as:

1. *Visual* or *astrometric* binaries, if both or one of the stars can be observed to move in a periodic fashion
2. *Spectrum* or *spectroscopic* binaries if there are one or two clearly identified spectra indicating different Doppler shifts. Spectroscopic binaries have sufficiently short orbital periods so that a changing Doppler shift can be measured
3. *Eclipsing* binaries if the light from the system is observed to vary periodically as each star is eclipsed by its companion

Note that a given binary can be placed in more than one of these classifications.

In principal, the masses of the components of a binary can be inferred from a measurement of its orbital properties.

2.1 The Two-Body Problem

Given a central force, the motion of two bodies is found from the Lagrangian, which can be expressed as

$$\mathscr{L} = \frac{1}{2}m_1 v_1^2 + \frac{1}{2}m_2 v_2^2 + \frac{Gm_1 m_2}{|\mathbf{r}_2 - \mathbf{r}_1|}. \tag{2.1}$$

Fig. 2.1 Barycenter coordinate description of a binary system

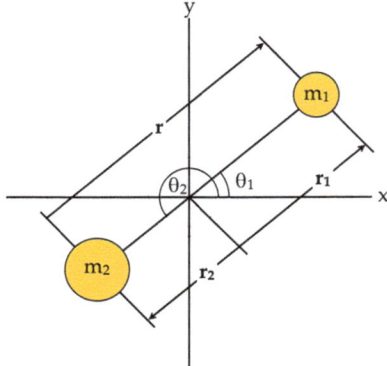

We choose a barycentric coordinate system, so that

$$m_1 \mathbf{r}_1 + m_2 \mathbf{r}_2 = 0 \qquad (2.2)$$

and therefore

$$m_1 r_1 = m_2 r_2. \qquad (2.3)$$

We define the relative separation to be

$$r = r_1 + r_2. \qquad (2.4)$$

We can use these two equations to solve for r_1 and r_2 in terms of r to get

$$r_1 = \frac{m_2}{M} r, \qquad (2.5)$$

$$r_2 = \frac{m_1}{M} r, \qquad (2.6)$$

where $M = m_1 + m_2$. Note that $\theta_1 = \theta_2 - \pi = \theta$ (Fig. 2.1).

Assuming that the orbits lie in a plane, we have

$$v_1^2 = \dot{r}_1^2 + r_1^2 \dot{\theta}_1^2 = \left(\frac{m_2}{M}\right)^2 \left(\dot{r}^2 + r^2 \dot{\theta}^2\right), \qquad (2.7)$$

$$v_2^2 = \dot{r}_2^2 + r_2^2 \dot{\theta}_2^2 = \left(\frac{m_1}{M}\right)^2 \left(\dot{r}^2 + r^2 \dot{\theta}^2\right) \qquad (2.8)$$

and so

$$\begin{aligned}
\mathscr{L} &= \frac{1}{2} \frac{m_1 m_2^2}{M^2} \left(\dot{r}^2 + r^2 \dot{\theta}^2\right) + \frac{1}{2} \frac{m_2 m_1^2}{M^2} \left(\dot{r}^2 + r^2 \dot{\theta}^2\right) + \frac{G m_1 m_2}{r} \\
&= \frac{1}{2} \frac{m_1 m_2}{M} \dot{r}^2 + \frac{1}{2} \frac{m_1 m_2}{M} r^2 \dot{\theta}^2 + \frac{G m_1 m_2 M}{M r} \\
&= \frac{1}{2} \mu \dot{r}^2 + \frac{1}{2} \mu r^2 \dot{\theta}^2 + \frac{G \mu M}{r}.
\end{aligned} \qquad (2.9)$$

2.2 The Orbital Shape

Problem 2.1: Demonstrate that the orbit lies in a plane by obtaining the Lagrangian using arbitrarily oriented spherical polar coordinates (r, ϕ, θ). Calculate the Euler–Lagrange equations of motion and show that one can recover the planar equations of motion using the initial conditions: $\theta = \pi/2$ and $\dot{\theta} = 0$.

Since $\mathscr{L}$ is independent of θ, we have

$$\frac{d}{dt}\frac{\partial \mathscr{L}}{\partial \dot{\theta}} - \frac{\partial \mathscr{L}}{\partial \theta} \Rightarrow \frac{\partial \mathscr{L}}{\partial \dot{\theta}} = \text{constant} \qquad (2.10)$$

so

$$\frac{\partial \mathscr{L}}{\partial \dot{\theta}} = \mu r^2 \dot{\theta} = J = \text{angular momentum}. \qquad (2.11)$$

The total energy is also conserved, and it is given by

$$\frac{1}{2}m_1 v_1^2 + \frac{1}{2}m_2 v_2^2 - \frac{Gm_1 m_2}{r} = \frac{1}{2}\mu\left(\dot{r}^2 + r^2 \dot{\theta}^2\right) - \frac{G\mu M}{r} = C. \qquad (2.12)$$

(Note that here we use C for the total energy instead of E—this is because E is reserved for the *eccentric anomaly*, which is an important quantity for describing observations of orbits.) Using Eq. (2.11), we can express the total energy as an equation that is dependent upon r only.

$$\dot{\theta} = \frac{J}{\mu r^2} \Rightarrow \dot{\theta}^2 = \frac{J^2}{\mu^2 r^4}, \qquad (2.13)$$

so

$$C = \frac{1}{2}\mu \dot{r}^2 + \frac{1}{2}\frac{J^2}{\mu r^2} - \frac{G\mu M}{r}. \qquad (2.14)$$

2.2 The Orbital Shape

From Eq. (2.14), we can obtain the time dependence of the radius of the orbit, and then we can obtain the time dependence of the orbital angle using Eq. (2.11). However, these results are not particularly useful for determining the orbit directly from observations of binaries. Instead, we will first determine the shape of the orbit using Eq. (2.14) and some clever variable substitutions. Later we will determine the time dependence of the orbit in terms of observational quantities.

In order to determine the shape of the orbit, we first make the variable substitution $u = 1/r$, so that

$$\frac{du}{d\theta} = u' = -\frac{1}{r^2}\frac{dr}{d\theta} \Rightarrow \frac{dr}{d\theta} = -r^2 u'. \qquad (2.15)$$

Now,
$$\dot{r} = \frac{dr}{d\theta}\dot{\theta} = -r^2 u' \frac{J}{\mu r^2} = -\frac{J}{\mu}u', \qquad (2.16)$$

where the $\dot{\theta}$ substitution comes from Eq. (2.13). Substitution of Eq. (2.16) into Eq. (2.14) gives

$$\frac{J^2}{2\mu}u'^2 + \frac{J^2}{2\mu}u^2 - G\mu M u = C. \qquad (2.17)$$

Now, we make another substitution and let $\ell = J^2/G\mu^2 M$ so that $J^2/\mu = GM\mu\ell$, and

$$\frac{1}{2}GM\mu\ell u'^2 + \frac{1}{2}GM\mu\ell u^2 - GM\mu u = C. \qquad (2.18)$$

Finally, we divide by $GM\mu/2\ell$ and add 1 to both sides to obtain

$$\ell^2 u'^2 + \ell^2 u^2 - 2\ell u + 1 = \frac{2C\ell}{GM\mu} + 1. \qquad (2.19)$$

Next, we define

$$e^2 = \frac{2C\ell}{GM\mu} + 1 \qquad (2.20)$$

and make the final substitution of $x = \ell u - 1$, so we have

$$x'^2 + x^2 = e^2, \qquad (2.21)$$

or

$$x' = \sqrt{e^2 - x^2}. \qquad (2.22)$$

This equation can be integrated as follows:

$$\int_{x_0}^{x} \frac{dx}{\sqrt{e^2 - x^2}} = \int_{\theta_0}^{\theta} d\theta,$$

$$\arcsin\left(\frac{x}{e}\right) - \arcsin\left(\frac{x_0}{e}\right) = \theta - \theta_0. \qquad (2.23)$$

Clearly, $|x| \leq |e|$ in order for the arcsin to make any sense. We define $\theta_0 = 0$ and require $x(0) = e$ to obtain

$$\arcsin\left(\frac{x(0)}{e}\right) - \arcsin\left(\frac{x_0}{e}\right) = 0 \Rightarrow \arcsin\left(\frac{x_0}{e}\right) = \arcsin 1 = \frac{\pi}{2}. \qquad (2.24)$$

Thus,

$$\frac{x}{e} = \sin(\theta + \pi/2) = \cos\theta \Rightarrow x = e\cos\theta. \qquad (2.25)$$

Reversing all the substitutions, we finally obtain

$$r = \frac{\ell}{(1 + e\cos\theta)}, \qquad (2.26)$$

2.2 The Orbital Shape

Fig. 2.2 Elliptical orbit with the origin centered on one star

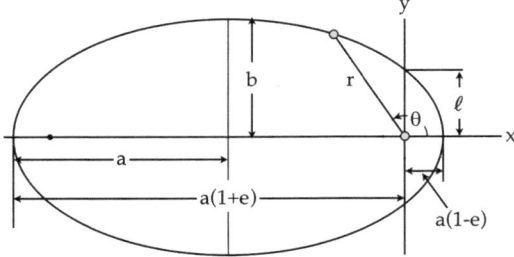

which is the parametric equation for an ellipse. Thus, the shape of the relative orbit is an ellipse with the point of closest approach (or *periastron*) at $\theta = 0$ and one body at the focus. The semimajor axis (a) of an ellipse is half of the long axis, which is also the sum of the minimum distance and the maximum distance (the *apastron*). Thus,

$$r_{\min} = r(0) = \ell/(1+e), \tag{2.27}$$
$$r_{\max} = r(\pi) = \ell/(1-e) \tag{2.28}$$

and

$$a = \frac{1}{2}(r_{\min} + r_{\max}) = \ell/(1-e^2) \Rightarrow \ell = a(1-e^2). \tag{2.29}$$

The periastron and apastron can now be expressed in terms of the semimajor axis as

$$r_{\min} = a(1-e), \tag{2.30}$$
$$r_{\max} = a(1+e). \tag{2.31}$$

Although initially introduced to simplify the differential equation, the value of e is found to be the *eccentricity* of the elliptical orbit (Fig. 2.2).

Problem 2.2: Derive Kepler's third law ($GM = a^3\omega^2$) using $J = \mu r^2 \dot\theta$ and $r = \ell/(1+e\cos\theta)$.

The actual motion of the components of the binary are about the center of mass (also known as the *barycenter*). We can show that this motion is also elliptical and obeys a version of Kepler's third law. Using barycentric coordinates, we have $m_1 \mathbf{r}_1 = -m_2 \mathbf{r}_2$ and $\mathbf{r}_1 - \mathbf{r}_2 = \mathbf{r}$. Therefore, from Newton's law, we have

$$\ddot{\mathbf{r}}_1 = -\frac{Gm_2}{r^3}\mathbf{r} = -\frac{Gm_2}{r_1^3}\left(\frac{m_2}{M}\right)^3 \mathbf{r}_1 \left(\frac{M}{m_2}\right) = -\frac{Gm_2^3}{M^2}\frac{\mathbf{r}_1}{r_1^3}. \tag{2.32}$$

We can obtain a similar equation for the motion of m_2 by simply interchanging 1 and 2. Note that these equations of motion are similar to the relative equation:

$$\ddot{\mathbf{r}} = -\frac{GM}{r^3}\mathbf{r}, \tag{2.33}$$

$$\ddot{\mathbf{r}}_1 = -\frac{G\left(m_2^3/M^2\right)}{r_1^3}\mathbf{r}_1, \tag{2.34}$$

$$\ddot{\mathbf{r}}_2 = -\frac{G\left(m_1^3/M^2\right)}{r_2^3}\mathbf{r}_2, \tag{2.35}$$

and so they all obey a version of Kepler's third law with the following values for the mass:

Relative: M
Barycentric body 1: m_2^3/M
Barycentric body 2: m_1^3/M

Note also that there is a simple rescaling of the position vectors between the barycentric frame and the relative orbit frame:

$$\mathbf{r}_1 = \frac{m_2}{M}\mathbf{r}. \tag{2.36}$$

$$\mathbf{r}_2 = -\frac{m_1}{M}\mathbf{r}, \tag{2.37}$$

and so the barycentric orbits are simply rescaled versions of the relative orbit ellipse.

2.3 Time-Dependent Orbits

The orbital shape of the barycentric orbits is of value when we can only observe one star in the binary system. If we see both stars and can identify the motion of the barycenter, then we can identify the individual masses of the stars. Frequently, we only measure part of the orbit, and often we only measure the orbital speed. Thus, we need to know the position of the components as a function of time. This is found from what is known as *Kepler's equation*. To derive this we need to study the geometry of an ellipse.

Consider an ellipse with semimajor axis a that is circumscribed by a circle of radius a, as shown in Fig. 2.3.

Referring to the figure, the following line segments and angles can be defined:

$$O\Pi = a = \text{semimajor axis}, \tag{2.38}$$

$$S\Pi = a(1-e) = \text{periastron}, \tag{2.39}$$

$$OS = ae \tag{2.40}$$

2.3 Time-Dependent Orbits

Fig. 2.3 Properties of an ellipse

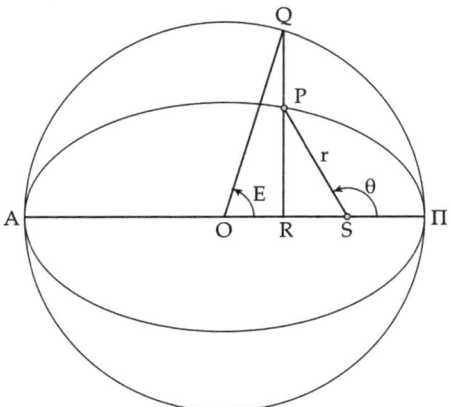

and

- The angle θ is called the *true anomaly*.
- The angle E is called the *eccentric anomaly*.

We want to find the time dependence of the eccentric anomaly, E.

The *auxiliary circle* has the property that $PR/QR = b/a = \sqrt{1-e^2}$, so

$$r\cos\theta = -RS = OS - OR = a\cos E - ae, \qquad (2.41)$$

$$r\sin\theta = PR = \left(\sqrt{1-e^2}\right)QR = a\sin E\sqrt{1-e^2}, \qquad (2.42)$$

and

$$\begin{aligned}
r &= \sqrt{r^2\cos^2\theta + r^2\sin^2\theta} \\
&= \sqrt{a^2e^2 - 2a^2e\cos E + a^2\cos^2 E + a^2\sin^2 E - a^2e^2\sin^2 E} \\
&= \sqrt{a^2e^2\left(1-\sin^2 E\right) - 2a^2e\cos E + a^2} \\
&= a\sqrt{e^2\cos^2 E - 2e\cos E + 1} \\
&= a(1-e\cos E). \qquad (2.43)
\end{aligned}$$

We use the equation for the *specific angular momentum*, or angular momentum per mass:

$$r^2 d\theta = L dt \qquad (2.44)$$

(n.b.: $L = J/\mu$), so we can substitute $r = a(1-e\cos E)$, but we still need an equation for θ.

We obtain this equation by noting that

$$\frac{d}{dE}\sin\theta = \cos\theta\frac{d\theta}{dE}. \qquad (2.45)$$

Using

$$\sin\theta = \frac{a}{r}\sin E\sqrt{1-e^2}$$

$$= \frac{a\sin E}{a(1-e\cos E)}\frac{b}{a}$$

$$= \frac{b\sin E}{a(1-e\cos E)} \tag{2.46}$$

and differentiating with respect to E gives

$$\frac{d}{dE}(\sin\theta) = \frac{b}{a}\frac{\cos E - e}{(1-e\cos E)^2} \tag{2.47}$$

so that

$$\cos\theta\, d\theta = \frac{b(\cos E - e)\, dE}{a(1-e\cos E)^2}. \tag{2.48}$$

Now, using

$$\cos\theta = \frac{-a(e-\cos E)}{r} = \frac{-(e-\cos E)}{(a-e\cos E)} \tag{2.49}$$

we find that

$$d\theta = -\frac{(1-e\cos E)}{(e-\cos E)}\frac{b(\cos E - e)}{a(1-e\cos E)^2}\, dE = \frac{b\, dE}{a(1-e\cos E)}. \tag{2.50}$$

Finally, we have

$$a^2(1-e\cos E)^2\frac{b\, dE}{a(1-e\cos E)} = L\, dt \tag{2.51}$$

or

$$(1-e\cos E)\, dE = \frac{L}{ab}\, dt. \tag{2.52}$$

Integrating this equation gives

$$\int (1-e\cos E)\, dE = \frac{L}{ab}\int dt \tag{2.53}$$

or

$$E - e\sin E = \frac{L}{ab}t + k. \tag{2.54}$$

Now we need to determine the integration constant. First, we define T to be the time at periastron passage and we note that $E = 0$ at periastron, so

$$k = -\frac{L}{ab}T \tag{2.55}$$

2.4 The Orbital Elements

and
$$E - e\sin E = \frac{L}{ab}(t - T). \tag{2.56}$$

From Kepler's second law, we have $\frac{1}{2}r^2 d\theta = dA = \frac{1}{2}L dt$, so

$$\int_0^{2\pi} \frac{1}{2}r^2 d\theta = \pi ab = \frac{1}{2}LP \Rightarrow \frac{L}{ab} = \frac{2\pi}{P} = \omega \tag{2.57}$$

and then
$$E - e\sin E = \frac{2\pi}{P}(t - T). \tag{2.58}$$

This equation is generally solved using numerical techniques. The simplest approach is to use a Newton–Raphson iterative solution—given x_{n-1}, we find x_n by

$$x_n = x_{n-1} - f(x_{n-1})/f'(x_{n-1}). \tag{2.59}$$

Here, we let $f(E) = E - e\sin E - 2\pi(t-T)/P$ and note that $f'(E) = 1 - e\cos E$.

2.4 The Orbital Elements

Observed binaries do not lie in the plane of the sky, so we need to describe the orientation of the binary using the *orbital elements*. These are defined in terms of both the total angular momentum vector **J** and the total energy of the orbit.

The orientation of the binary can be described in terms of the direction of the total angular momentum vector and the direction of the periastron, which give the z- and x-axes in the orbital plane, respectively. These directions are measured relative to a coordinate system that is defined by the tangent plane to the celestial sphere at the location of the binary. A Cartesian coordinate system is defined in terms of the line of sight to the binary from the observer and the tangent to a great circle joining the binary to the north celestial pole. The angle of inclination is defined as the angle between the plane of the orbit and the tangent plane to the celestial sphere. The ascending node (N) is the line defined by the intersection of the plane of the orbit and the tangent plane and points in the direction where the binary passes from *inside* the celestial sphere to *outside* the celestial sphere. Figure 2.4 shows the orientation of the orbit relative to the tangent plane and the three angles that define this orientation. These three angles are

Angle of inclination i
Longitude of the ascending node Ω
Longitude of the periastron ω

The shape of the orbit is then given by three quantities:

Semimajor axis a
Eccentricity e
Time of periastron T

Fig. 2.4 Illustration of the different angles for the orbital elements of a binary. The line of sight from the observer to the binary is the along the z-axis, viewed from $z = -\infty$. The x-axis is chosen so that the positive x direction points toward the north celestial pole. The tangent plane of the sky is the xy-plane

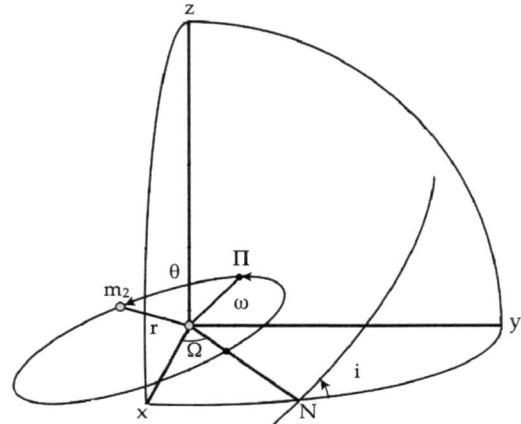

These six quantities are called the orbital elements. If the orbital elements can be measured, then the masses of the binary can be determined. The orbit will always appear to be an ellipse when viewed on the sky, but unless $i = 0$, the center of mass of the system will not lie at the focus of this apparent ellipse (Fig. 2.5).

The angular momentum and total energy are also related to the orbital period and orbital shape. To obtain these relations we begin by noting that the kinetic energy is

$$K = \frac{1}{2}m_1 v_1^2 + \frac{1}{2}m_2 v_2^2 = \frac{1}{2}\mu v^2, \qquad (2.60)$$

where $v^2 = \dot{r}^2 + r^2 \dot{\theta}^2$ and r and θ are relative separation variables. Now, using $r = \ell/(1 + e\cos\theta)$, we find that

$$\dot{r} = \dot{\theta}\frac{r^2}{\ell} e \sin\theta = \frac{L}{\ell} e \sin\theta \qquad (2.61)$$

and

$$r\dot{\theta} = \frac{r^2 \dot{\theta}}{r} = \frac{L}{r} = \frac{L}{\ell}(1 + e\cos\theta). \qquad (2.62)$$

From here we get

$$\begin{aligned}
v^2 &= \left(\frac{L}{\ell}\right)^2 \left[e^2 \sin^2\theta + 1 + 2e\cos\theta + e^2 \cos^2\theta\right] \\
&= \left(\frac{L}{\ell}\right)^2 \left[e^2 + 1 + 2e\cos\theta\right] = \left(\frac{L}{\ell}\right)^2 \left[2(1 + \cos\theta) - 1 + e^2\right] \\
&= \left(\frac{L}{\ell}\right)^2 \left[\frac{2\ell}{r} - (1 - e^2)\right] = \frac{L^2}{\ell}\left[\frac{\ell}{r} - \frac{1 - e^2}{\ell}\right] \\
&= \frac{L^2}{\ell}\left[\frac{2}{r} - \frac{1}{a}\right],
\end{aligned} \qquad (2.63)$$

2.4 The Orbital Elements

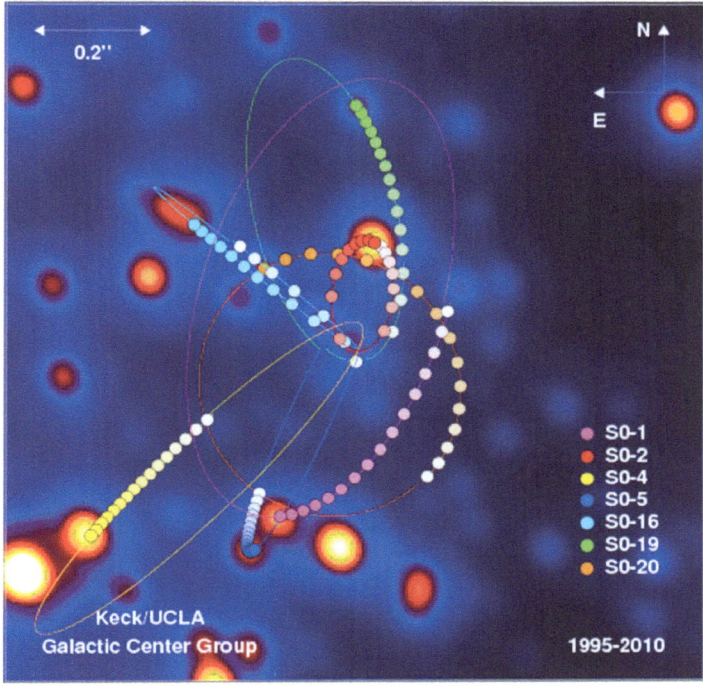

Fig. 2.5 Orbits of stars around Sgr A*. Note that every orbit is an ellipse, but that the foci do not all lie at a common point, even though all orbits are about the same object. This image was created by Prof. Andrea Ghez and her research team at UCLA and is from data sets obtained with the W. M. Keck Telescopes

where we have used $a = \ell\left(1-e^2\right)$ in the last step. Now from Kepler's second law, we have $L = 2\pi ab/P$, where P is the orbital period. Noting that $b^2 = a^2(1-e^2)$ we find

$$\begin{aligned} L &= \frac{4\pi^2 a^2 b^2}{P^2} = \frac{4\pi^2 a^3}{P} a\left(1-e^2\right) \\ &= GMa\left(1-e^2\right) \\ &= GM\ell, \end{aligned} \tag{2.64}$$

where we have used Kepler's third law. Finally, we have

$$v^2 = GM\left[\frac{2}{r} - \frac{1}{a}\right], \tag{2.65}$$

and so the kinetic energy is

$$K = \frac{1}{2}\mu v^2 = \frac{1}{2}\frac{m_1 m_2}{M}GM\left[\frac{2}{r} - \frac{1}{a}\right]$$

$$= \frac{Gm_1 m_2}{r} - \frac{Gm_1 m_2}{2a}. \quad (2.66)$$

Now, the potential energy is $\Omega = -Gm_1 m_2/r$, so the total energy is

$$C = K + \Omega = -\frac{Gm_1 m_2}{2a}. \quad (2.67)$$

The total angular momentum is $J = m_1 L_1 + m_2 L_2$, where

$$L_1 = \frac{m_2^2}{M^2}L, \quad (2.68)$$

$$L_2 = \frac{m_1^2}{M^2}L, \quad (2.69)$$

$$L^2 = GMa\left(1 - e^2\right). \quad (2.70)$$

This gives:

$$J = \frac{1}{M^2}\left(m_1 m_2^2 + m_2 m_1^2\right)\sqrt{GMa(1-e^2)}$$

$$= m_1 m_2 \sqrt{\frac{Ga(1-e^2)}{M}}$$

$$= \frac{2\pi}{P}\frac{m_1 m_2 a^2 \sqrt{1-e^2}}{M}. \quad (2.71)$$

Thus, the total energy is fixed by the masses and the semimajor axis, while the total angular momentum also depends upon the period and the eccentricity.

2.5 Spectroscopic Binaries

We now look at determining the mass from spectroscopic binaries, where we can only measure the radial velocity of the component stars. The Doppler shift alters the frequency of spectral lines in stars by

$$f' = f\sqrt{\frac{c \pm v}{c \mp v}}, \quad (2.72)$$

where v is the radial velocity of the star and the sign choice depends on whether the star is moving toward us or away from us. From the frequency shifts, we can

2.5 Spectroscopic Binaries

determine the total radial velocity which is a combination of the systemic motion of the binary and the velocity of the individual stars, so

$$v_{\text{rad}} = \dot{z} + \gamma. \tag{2.73}$$

From the orbital elements, we see that the z-component of the star in its orbit is given by

$$z = r \sin(\theta + \omega) \sin i, \tag{2.74}$$

and so the radial velocity is

$$\dot{z} = \sin i \left[\dot{r} \sin(\theta + \omega) + r\dot{\theta} \cos(\theta + \omega) \right]. \tag{2.75}$$

Since $r = a(1-e^2)/(1+e\cos\theta)$, we have

$$\dot{r} = er\dot{\theta} \sin\theta / (1 + e\cos\theta). \tag{2.76}$$

Also, we have $r^2\dot{\theta} = 2\pi a^2 \sqrt{1-e^2}/P$, and so

$$r\dot{\theta} = 2\pi a^2 \sqrt{1-e^2}/rP = \frac{2\pi a(1+e\cos\theta)}{P\sqrt{1-e^2}}. \tag{2.77}$$

Substituting these two equations into Eq. (2.75), we find

$$\dot{z} = \frac{2\pi a \sin i}{P\sqrt{1-e^2}} \left[\cos(\theta + \omega) + e\cos\omega \right], \tag{2.78}$$

and so the total measured radial velocity is

$$v_{\text{rad}} = K \left[\cos(\theta + \omega) + e\cos\omega \right] + \gamma, \tag{2.79}$$

where $K = (2\pi a \sin i)/\left(P\sqrt{1-e^2}\right)$ is the *semi-amplitude of the velocity* and γ is the radial velocity of the center of mass. Note that K is not to be confused with the kinetic energy described in the previous section. A remarkable consequence of this result is that the extrema of v_{rad} are at the line of nodes. Several velocity curves for a variety of binary systems are shown in Fig. 2.6.

We can determine the value of K observationally by measuring the maximum and minimum velocities through the Doppler shift of spectral lines. Note that these values occur at $\theta + \omega = 0$ and π, respectively. Therefore,

$$v_{\max} = K[e\cos\omega + 1] + \gamma, \tag{2.80}$$

$$v_{\min} = K[e\cos\omega - 1] + \gamma, \tag{2.81}$$

and so

$$v_{\max} - v_{\min} = 2K \tag{2.82}$$

or

$$K = \frac{1}{2}(v_{\max} - v_{\min}). \tag{2.83}$$

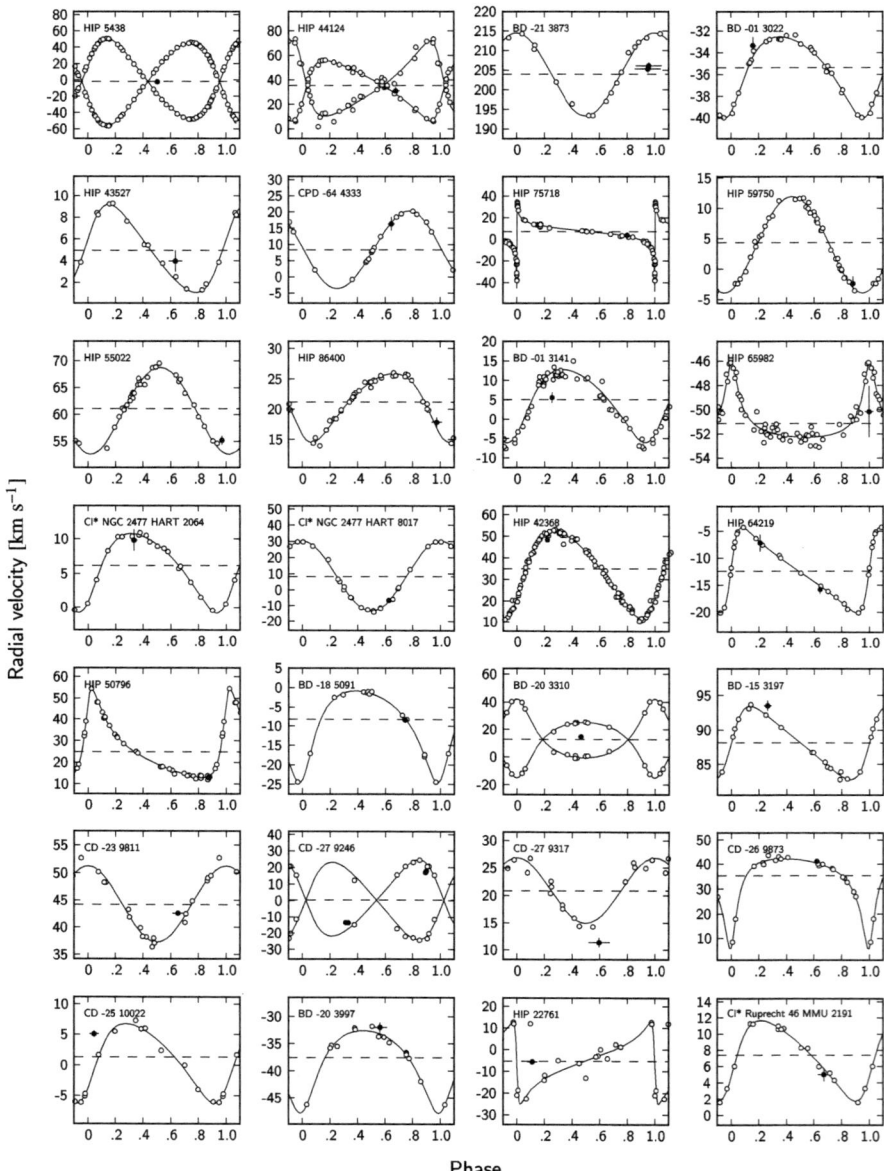

Fig. 2.6 Various velocity curves for several binary systems. Some are single-lined and some are double-lined. Figure taken from Matijevič, et al., *Astron. J.*, 141, 200 (2011). Reproduced by permission from the AAS

2.5 Spectroscopic Binaries

By fitting Eq. (2.75) to the shape of the velocity curve, one can obtain the values of e, ω, and γ. If a double-lined spectroscopic binary is observed, then we can determine

$$K_1 = \frac{2\pi a_1 \sin i}{P\sqrt{1-e^2}}, \tag{2.84}$$

$$K_2 = \frac{2\pi a_2 \sin i}{P\sqrt{1-e^2}} \tag{2.85}$$

along with e, ω, and γ. Therefore, we know

$$a_1 \sin i = \frac{\sqrt{1-e^2}}{2\pi} K_1 P, \tag{2.86}$$

$$a_2 \sin i = \frac{\sqrt{1-e^2}}{2\pi} K_2 P. \tag{2.87}$$

Since we know $m_1 a_1 = m_2 a_2$ and $GM = 4\pi^2 a^3/P^2$, we make the substitution:

$$m_2 = m_1 (a_1/a_2) = m_1 \left(\frac{a_1 \sin i}{a_2 \sin i}\right) = m_1 (K_1/K_2) \tag{2.88}$$

so

$$Gm_1 \left(1 + \frac{K_1}{K_2}\right) = \frac{4\pi^2}{P^2} (a_1 \sin i + a_2 \sin i)^3 / \sin^3 i \tag{2.89}$$

and

$$m_1 \sin^3 i = \frac{4\pi^2}{P^2} \frac{K_2}{G(K_1+K_2)} \left(\frac{\sqrt{1-e^2}}{2\pi} P\right)^3 (K_1+K_2)^3$$

$$= \frac{P}{2\pi G} (1-e^2)^{3/2} (K_1+K_2)^2 K_2. \tag{2.90}$$

This provides an upper limit for m_1 unless i is known. We can find an upper limit for m_2 by simply interchanging 1 and 2 in Eq. (2.90). If we can only measure the radial velocity of one component of the binary (say K_1), then we can determine the *mass function* by using Eq. (2.88) to determine K_2 in terms of m_1, m_2, and K_1. We substitute this expression for K_2 into Eq. (2.90) to obtain

$$m_2 \sin^3 i = \frac{P K_1^3}{2\pi G} (1-e^2)^{3/2} \left(\frac{m_1+m_2}{m_2}\right)^2, \tag{2.91}$$

and so

$$f(m) = \frac{m_2^3 \sin^3 i}{(m_1+m_2)^2} = \frac{P K_1^3}{2\pi G} (1-e^2)^{3/2}, \tag{2.92}$$

where $f(m)$ is known as the mass function.

If the orbit is also a visual binary, it is possible to obtain the angle of inclination and consequently to obtain exact values for m_1 and m_2. The direct measurement of the masses of all stars except the sun is determined in this way.

> **Problem 2.3:** MT720 is a spectroscopic binary in the Cygnus OB2 Association. It is found to have a period of $P = 4.36$ d and an eccentricity of e $= 0.35$. The semi-amplitude of the radial velocities are $K_1 = 173 \text{ km/s}$ and $K_2 = 242 \text{ km/s}$.
>
> (a) Find $m \sin^3 i$ and $a \sin i$ for each star.
> (b) What is the mass ratio: $q = m_2/m_1$?
> (c) If $i = 70°$, what are the masses of each star?

Problems

2.1. Demonstrate that the orbit lies in a plane by obtaining the Lagrangian using arbitrarily oriented spherical polar coordinates (r, ϕ, θ). Calculate the Euler–Lagrange equations of motion and show that one can recover the planar equations of motion using the initial conditions: $\theta = \pi/2$ and $\dot{\theta} = 0$.

2.2. Derive Kepler's third law $(GM = a^3 \omega^2)$ using $J = \mu r^2 \dot{\theta}$ and $r = \ell/(1 + e\cos\theta)$.

2.3. MT720 is a spectroscopic binary in the Cygnus OB2 Association. It is found to have a period of $P = 4.36$ d and an eccentricity of e $= 0.35$. The semi-amplitude of the radial velocities are $K_1 = 173 \text{ km/s}$ and $K_2 = 242 \text{ km/s}$.

(a) Find $m \sin^3 i$ and $a \sin i$ for each star.
(b) What is the mass ratio: $q = m_2/m_1$?
(c) If $i = 70°$, what are the masses of each star?

Chapter 3
Measuring Other Stellar Properties

The distance to nearby stars can be measured through geometrical methods. When the distance is known, additional observations of the spectra of stars and light curves of binary systems allow us to determine the temperatures, sizes, and luminosities of many nearby stars. In this way we can find correlations between different stellar properties and begin to sort stars into different categories and classifications.

3.1 Distances and Parallax

If their distances can be measured, we can find the luminosity of stars through their magnitudes. Although there are many methods for approximating the distance to stars (also known as the distance ladder), they are all based on the only direct measurement method currently known—parallax.

Parallax is a triangulation method using the diameter of the earth's orbit as a baseline. In this method, the celestial coordinates of a star are measured at two times separated by six months. If the star is seen to shift by an angle 2α during this interval, we can construct an isosceles triangle whose short side is the diameter of the earth's orbit about the sun and whose equal sides are the distance to the star as shown in Fig. 3.1. The distance can be expressed in terms of the *astronomical unit* (AU), which is the mean distance between the earth and the sun and has a value of 149.6×10^9 m. Because the distance to stars is so great, only nearby stars have an angular shift in apparent sky location that is large enough to be measurable. These angular shifts (α) are all less than one arcsecond. The standard unit of distance measurement used for stellar distances is the *parsec* (pc), which is defined to be the distance at which one astronomical unit subtends an arcsecond. Since the angular shift decreases with increasing distance, we define the distance in pc to be

$$d \text{ pc} = 1/\alpha, \tag{3.1}$$

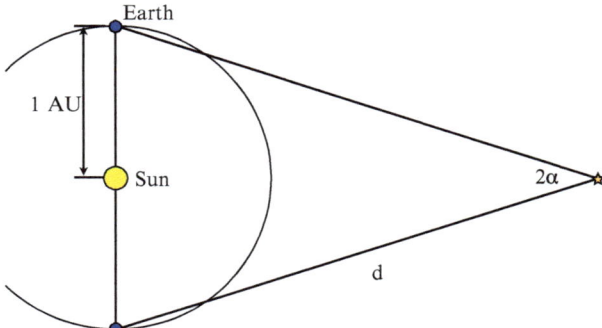

Fig. 3.1 Geometry of a parallax measurement. In reality, d is substantially larger and α is substantially smaller than shown in the figure

where α is measured in arcseconds. (Tables of the nearest and brightest stars can be found in Appendix C.)

We measure the apparent brightness of stars in terms of the intensity of light striking our instruments, so that $I =$ Power/Area. The intensity is related to the total luminosity by

$$I = \frac{L}{4\pi d^2}, \tag{3.2}$$

so the luminosity can be determined from a measurement of I and d (i.e., $L = I 4\pi d^2$). This allows us to measure the power output of stars. Some of these stars are also in binary systems, so that their masses can also be known. If we know two intrinsic properties of stars, it is instructive to see if there is a correlation between these properties. It turns out that there is a relationship between the mass and the luminosity of a star. The *mass–luminosity* relation is a power law:

$$L \propto M^\nu, \tag{3.3}$$

where ν appears to be around 3.5 for stars between about 1 $M_\odot$ and 20 $M_\odot$. It flattens out to $\nu \sim 1$ for very high mass stars. Recent data for intermediate-mass stars is shown in Fig. 3.2.

3.2 Temperature and Blackbody Spectrum

If the spectrum of a star can also be measured, the first thing we notice is that the bulk shape of the spectrum is a blackbody spectrum with absorption lines (although some have emission lines as well) and some nonthermal features (Fig. 3.3). A blackbody spectrum arises from the distribution of thermal energy in photons, given that the energy per photon is given by $E = h\nu$. We can obtain an expression for the spectrum by calculating the energy density of photons as a function of frequency ($u(\nu)$) for a gas of temperature T:

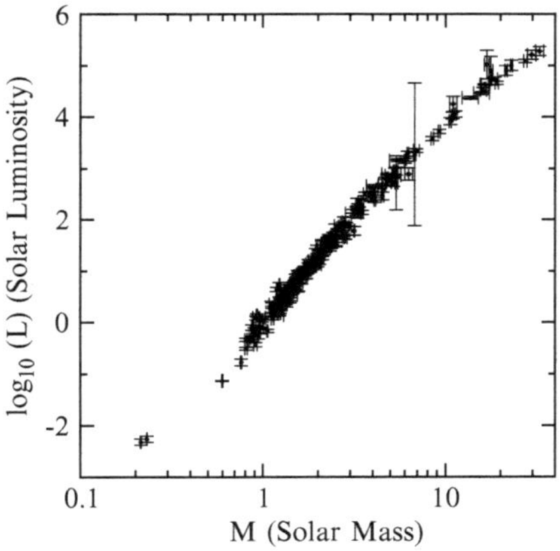

3.2 Temperature and Blackbody Spectrum

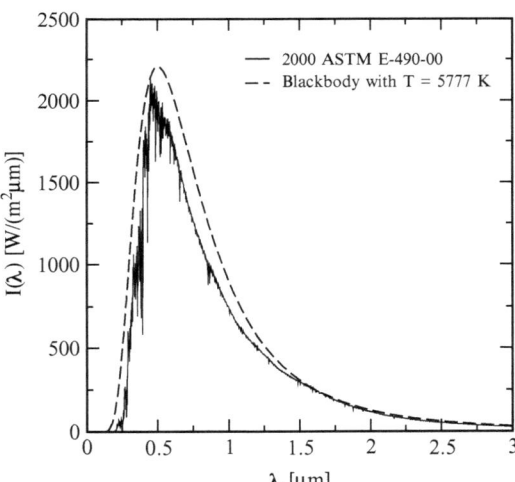

Fig. 3.2 The masses and luminosity of 201 intermediate-mass main sequence stars showing the power-law relationship between the two quantities. This plot was generated using data from O. Yu. Malkov, *Monthly Notices of the Royal Astronomical Society*, Vol. 382, p. 1073 (2007)

Fig. 3.3 Standard solar reference spectrum, compared with a pure blackbody spectrum calculated for the effective temperature of the sun, $T_{\text{eff}} = 5777$ K. This data is used with permission from the National Renewable Energy Laboratory (NREL), http://rredc.nrel.gov/solar/spectra/am0/E490_00a_AM0.xls. The ASTM standard is from ASTM Standard C33, 2003, "Specification for Concrete Aggregates," ASTM International, West Conshohocken, PA, 2003, DOI: 10.1520/C0033-03, www.astm.org

$$u(\nu)d\nu = \bar{E}(\nu)g(\nu)d\nu, \qquad (3.4)$$

where $\bar{E}(\nu)$ is the average energy per state of frequency ν, and $g(\nu)d\nu$ is the density of states with frequency ν per volume. The units of $u(\nu)$ are $J \cdot s/m^3$.

We can find the average energy per state by noting that the probability that a given state will have energy E for a given temperature is

$$f(E) = Ce^{-E/kT}, \qquad (3.5)$$

where C is a normalization constant to be determined later. Since photons come in discrete energies, this is also the probability that a given state will contain n photons, and can be expressed as

$$f_n = Ce^{-nh\nu/kT}. \qquad (3.6)$$

Fig. 3.4 Positive octant of a spherical shell in m-space with radius m and thickness dm

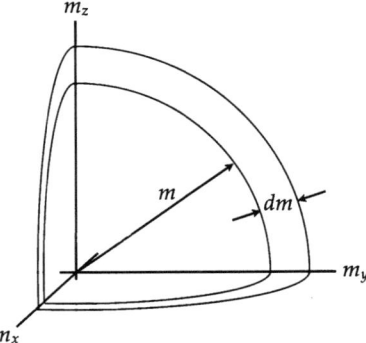

The normalization constant can now be found by requiring $\sum_{n=0}^{\infty} f_n = 1$, or

$$1 = C \sum_{n=0}^{\infty} \left(e^{-h\nu/kT}\right)^n = \frac{C}{1 - e^{-h\nu/kT}} \implies C = 1 - e^{-\beta h\nu}, \tag{3.7}$$

where $\beta = 1/kT$. From here, we can find the average energy per state through

$$\bar{E} = \sum_{n=0}^{\infty} E_n f_n = \sum_{n=0}^{\infty} nh\nu C e^{-n\beta h\nu} = -C \sum_{n=0}^{\infty} \frac{\partial}{\partial \beta} e^{-n\beta h\nu} = -C \frac{\partial}{\partial \beta} \sum_{n=0}^{\infty} e^{-n\beta h\nu}. \tag{3.8}$$

Therefore, from Eq. (3.7),

$$\bar{E} = -\left(1 - e^{-\beta h\nu}\right) \frac{\partial}{\partial \beta} \left(1 - e^{-\beta h\nu}\right)^{-1} = \frac{h\nu e^{-\beta h\nu}}{1 - e^{-\beta h\nu}}. \tag{3.9}$$

The density of states can be found by considering the solution to the wave equation for electromagnetic waves in a small cube of volume L^3. The boundary condition that the waves vanish at the edges of the cube results in the following conditions along each axis of the cube:

$$k_x L = m_x \pi, \quad k_y L = m_y \pi, \quad k_z L = m_z \pi, \tag{3.10}$$

where the m's are positive integers. This results in the following constraint on the wave number:

$$k = \frac{2\pi}{\lambda} = \sqrt{k_x^2 + k_y^2 + k_z^2} = \frac{\pi}{L} \sqrt{m_x^2 + m_y^2 + m_z^2} = \frac{\pi}{L} m. \tag{3.11}$$

Therefore, the number of states in a volume L^3, with energy between $h\nu$ and $h(\nu + d\nu)$ where $h\nu = hc/\lambda$ is equal to the number of ways you can choose the integers

3.2 Temperature and Blackbody Spectrum

m_x, m_y, and m_z so that they all add up to the same value of $m = \sqrt{m_x^2 + m_y^2 + m_z^2}$. This is equivalent to determining the volume of the positive octant of a spherical shell of radius m and thickness dm (Fig. 3.4). In addition, there are two spin states allowed for each photon with a given wave number, so the number of states is

$$2\left(\frac{1}{8}\right) 4\pi m^2 dm = \pi m^2 dm. \tag{3.12}$$

Now, since

$$h\nu = \frac{hc}{\lambda} = \frac{hc}{2L}m, \tag{3.13}$$

we have

$$m = \frac{2L\nu}{c}, \tag{3.14}$$

$$dm = \frac{2L}{c}d\nu, \tag{3.15}$$

and the number of states

$$g(\nu)d\nu = \frac{\pi m^2 dm}{L^3} = \frac{1}{L^3}\pi\left(\frac{2L\nu}{c}\right)^2\left(\frac{2L}{c}\right)d\nu = \frac{8\pi \nu^2}{c^3}d\nu. \tag{3.16}$$

Finally, we can combine Eq. (3.16) with Eq. (3.9) to obtain the blackbody spectrum:

$$u(\nu)d\nu = \frac{h\nu e^{-h\nu/kT}}{1 - e^{-h\nu/kT}}\frac{8\pi \nu^2}{c^3}d\nu = \frac{8\pi h\nu^3 d\nu}{c^3\left(e^{h\nu/kT} - 1\right)}. \tag{3.17}$$

From this, we can also obtain the number density of photons with a given frequency ν using

$$n(\nu)d\nu = \frac{u(\nu)d\nu}{h\nu} = \frac{8\pi \nu^2 d\nu}{c^3\left(e^{h\nu/kT} - 1\right)}. \tag{3.18}$$

Finally, using $\nu = c/\lambda$, we can recast the energy density in terms of wavelength:

$$u(\lambda)d\lambda = \frac{8\pi hc\, d\lambda}{\lambda^5\left(e^{hc/\lambda kT} - 1\right)}. \tag{3.19}$$

With this, we can determine $\lambda_{\max}$, the wavelength at which peak power is emitted. The solution to $du/d\lambda = 0$ can be found numerically, and the result is

$$\lambda_{\max} T = \text{constant} = 0.002897755 \text{ m} \cdot \text{K}. \tag{3.20}$$

By measuring $\lambda_{\max}$, we can obtain the effective temperature:

$$T_{\text{eff}} = \frac{0.0029 \text{ m} \cdot \text{K}}{\lambda_{\max}}. \tag{3.21}$$

The effective temperature is the temperature of the *photosphere*, the surface at which the atmosphere of the star becomes opaque. Below this surface, photons are scattered by the stellar atmosphere and their properties can be altered. Another way of looking at this surface is to think of it as the depth at which a photon traveling into the star has a 100% chance of being absorbed.

The temperature of the stars can be plotted against the luminosity to obtain the *Hertzsprung–Russell diagram* (or H-R diagram) which demonstrates a relation between these two quantities for most observed stars (see Fig. 3.5). Where it is not possible (or at least not practical) to measure the temperature and luminosity of stars, the "color" which is the difference in brightness of the star in two frequency bands can be plotted against the total brightness in order to produce a plot that is very similar to the Hertzsprung–Russell diagram. Such diagrams are called *color-magnitude diagrams* or CMDs.

If we assume that the star is a perfect blackbody we can estimate the stellar radius using the luminosity and the effective temperature. From the Stefan–Boltzmann law, we have

$$\sigma T_{\text{eff}}^4 = \frac{L}{4\pi R^2}, \tag{3.22}$$

where Stefan's constant is

$$\sigma = \frac{2\pi^5 k^4}{15 c^2 h^3} = 5.67 \times 10^{-8} \text{W}/\text{m}^2 \cdot \text{K}^4 \tag{3.23}$$

so

$$R = \sqrt{\frac{L}{4\pi \sigma T_{\text{eff}}^4}}. \tag{3.24}$$

Equation (3.22) can also be written as

$$L = 4\pi R^2 \sigma T_{\text{eff}}^4, \tag{3.25}$$

which shows that stars with large radii lie in the upper right of the H-R diagram and small stars lie in the lower left. This is reflected in the names of different classifications of stars shown in Fig. 3.5.

3.3 Radii and Eclipsing Binaries

The indirect measurement of stellar radii described in the preceding section can be confirmed using eclipsing binaries. The plot of the light from a star or stellar system as a function of time is known as the *light curve*. The light curve of an eclipsing binary shows two distinct dips as each star eclipses its companion. In order for eclipses to occur, the angle of inclination must be close to 90°. An idealized light curve is shown in Fig. 3.6.

3.3 Radii and Eclipsing Binaries

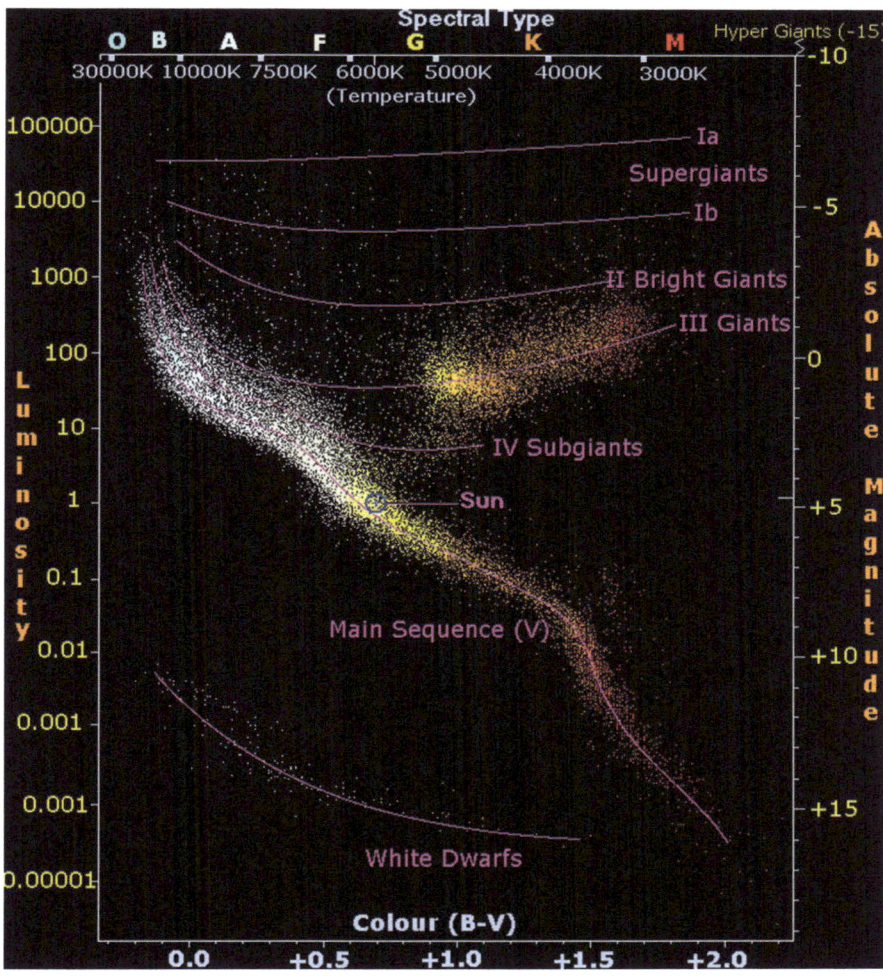

Fig. 3.5 Hertzsprung–Russell diagram. Figure produced by Richard Powell and available from http://www.atlasoftheuniverse.com/hr.html

On the one hand, if no radial velocity information is available, then the ratio of the radii of both stars can be found. If the timing of the eclipses does not occur at orbital phases separated by exactly one half of an orbit, then the orbit is eccentric and it may be possible to measure the eccentricity e and the argument of the periastron ω. On the other hand, if the radial velocity can be determined through analysis of the spectrum of one or both of the stars, then the techniques of Sect. 2.5 can be used to determine the orbital elements and orbital velocities of the system. In this case, the timing and duration of the eclipses can be used to obtain the absolute radii of the

Fig. 3.6 Geometry of an eclipsing system and its associated light curve

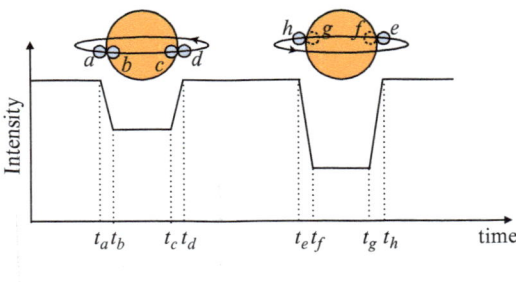

components. Eccentricity merely adds complications to the calculation of the stellar radii, so here we assume circular orbits in order to demonstrate the procedure. The brighter star is called the primary, while the other star is the secondary. Note that it is not necessarily the case that the primary is also the larger star. In this example, we assume that the primary is the smaller star.

Because the orbits are circular, the orbital speed of each component is constant. Thus the primary has speed v_p and the secondary has speed v_s. Since the inclination for an eclipsing binary is nearly $90°$, the maximum radial velocity is approximately the velocity of the star across the line of sight to the binary during the eclipses. Referring to Fig. 3.6, the time interval $t_b - t_a$ is the time it takes for the primary to cross the edge (or *limb*) of the secondary. Therefore, the distance traveled by the primary is equal to its diameter, and the radius of the primary is

$$r_p = \frac{v_p + v_s}{2}(t_b - t_a). \tag{3.26}$$

Similarly, the primary travels a distance equal to the diameter of the secondary during the time interval $t_c - t_a$, and so the radius of the secondary is

$$r_s = \frac{v_p + v_s}{2}(t_c - t_a). \tag{3.27}$$

Although the star being eclipsed is different, the same argument holds for the time intervals $t_f - t_e$ and $t_g - t_e$.

Additional complications arise when we consider that the surface brightness of the disk of a star is reduced near its edge (a phenomenon known as *limb darkening*). This results in more gradual transitions in the light curve during eclipses. Furthermore, if the stars are close enough to each other to tidally distort their shapes away from spherical, then the surface area of each star changes with the varying orientation toward our line of sight, adding more variation to the light curve. Finally, eccentricity and stellar rotations that are not synchronized to the orbital period can introduce variation in the light curve that are not symmetric about each eclipse. All of these effects must be considered when fitting an orbital and stellar model to eclipsing binary light curves in order to obtain the orbital elements and stellar radii of the system. Examples of several light curves are shown in Fig. 3.7

3.3 Radii and Eclipsing Binaries

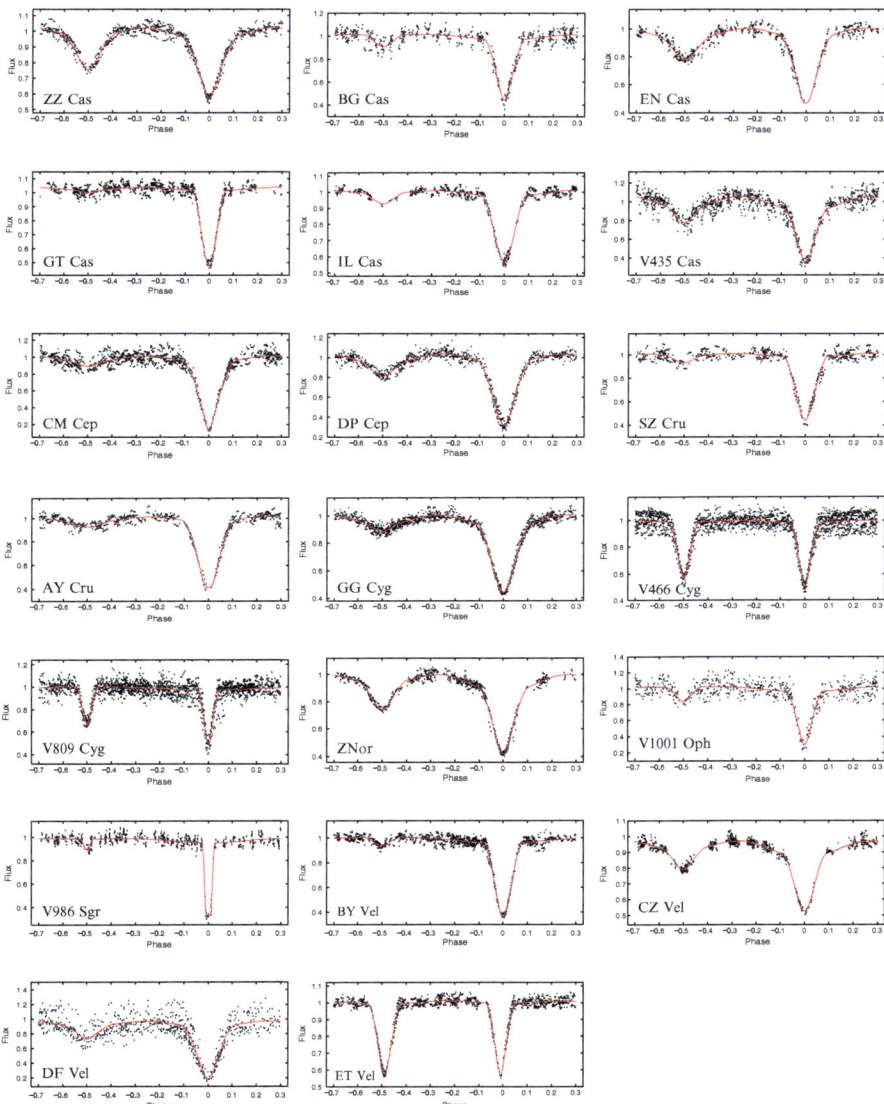

Fig. 3.7 Light curves for several binary systems. The *vertical axes* are normalized flux and the *horizontal axes* represent the orbital phase. Data from INTEGRAL/OMC. Reprinted from P. Zasche, *New Astronomy* 14, 129 (2009) with permission from Elsevier

Problem 3.1: GK Vir is an eclipsing spectroscopic binary with an angle of inclination $i = 89.5° \pm 0.6°$, so that it can be considered to be viewed edge on. The orbit is circular with semi-amplitudes of the radial velocities given by $K_1 = 38.6\,\text{km/s}$ and $K_2 = 221.6\,\text{km/s}$. The orbital period is $P = 0.344$ d. The time required for the light curve to drop to its lowest value is $t_b - t_a = 89.6\,\text{s}$, while the time required for the light curve to begin rising again is $t_c - t_a = 817\,\text{s}$. Use this information to find:

(a) The radii of both stars
(b) The masses of both stars

3.4 Boltzmann and Saha Equations

We can also classify stars by their absorption spectra. At the surface of stars (by this we mean the photosphere) we see absorption spectra when the gas near the surface absorbs light at wavelengths that are equal to the transition frequencies of the gas atoms. The dark lines of the absorption spectra arise because the likelihood of a photon scattering off of the gas at these frequencies is higher than at surrounding frequencies and so the surface of last scattering is somewhat higher in the atmosphere of the star where it is cooler and the luminosity is lower. The spectrum of the sun, showing numerous absorption lines is shown in Fig. 3.8.

Originally, stars were classified alphabetically according to the strength of the Lyman series of hydrogen lines. It was later shown that the relative strength of the Lyman lines is related to temperature, but not in a monotonic way. Consequently, when arranged according to decreasing temperature, the alphabetical ordering is lost. The sequence is now given by O, B, A, F, G, K, M. The spectra of different spectroscopic types are shown in Fig. 3.9.

The relative strength of these lines is related to the composition and temperature of the outer stellar atmosphere. This relationship is described by using a combination of the Boltzmann equation and the Saha equation. The Boltzmann equation describes the relative number of atoms in different excited energy levels. The Saha equation describes the relative number of atoms in different ionization states. In local thermal equilibrium, the amount of energy in excited atoms should be in equipartition with the kinetic energy of the atoms. Consequently, the number of atoms that are in the required excited state to absorb a photon is related to the temperature of the gas.

The atoms in the stellar atmosphere have a distribution of speeds described by the Maxwell–Boltzmann distribution:

$$n_v dv = n \left(\frac{m}{2\pi kT} \right)^{3/2} e^{-mv^2/2kT} 4\pi v^2 dv, \tag{3.28}$$

3.4 Boltzmann and Saha Equations

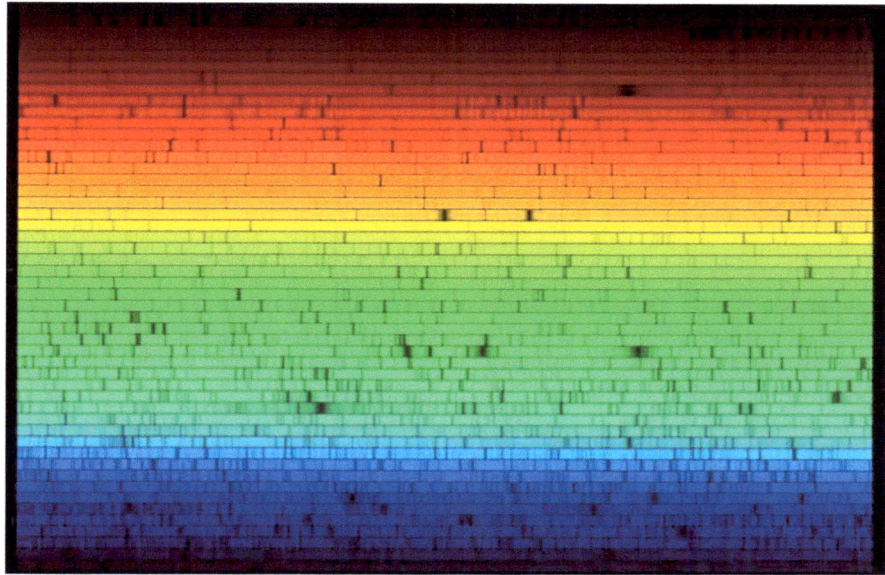

Fig. 3.8 Detailed solar spectrum, showing absorption lines. Figure courtesy of N.A. Sharp and National Optical Astronomy Observatory/Association of Universities for Research in Astronomy/National Science Foundation

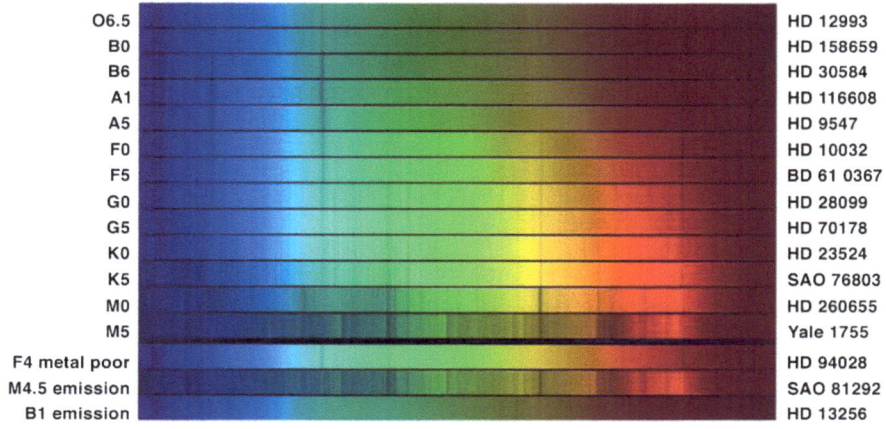

Fig. 3.9 Different stellar spectra, showing the varying intensities of the spectral lines with spectroscopic type. Figure courtesy of National Optical Astronomy Observatory/Association of Universities for Research in Astronomy/National Science Foundation

where n_v is the number density of atoms with speed v, n is the total number density of atoms in the gas, and m is the mass of the atoms. Individual atoms gain and lose energy through collisions, and so the distribution in kinetic energies of the atoms produces a characteristic distribution in excitation levels in the atoms. Let s_a be the specific set of quantum numbers associated with a state of energy E_a, and let s_b be associated with energy E_b. The ratio of the probability that the system is in state b relative to state a is

$$\frac{P(s_b)}{P(s_a)} = e^{-(E_b - E_a)/kT}. \tag{3.29}$$

If there are several states with different sets of quantum numbers, but the same energy, these states are said to be *degenerate*. If there are degenerate states (and there are), then we introduce *statistical weights* g_a and g_b which are the number of states of each energy, and the ratio of probabilities becomes

$$\frac{P(E_b)}{P(E_a)} = \frac{g_b}{g_a} e^{-(E_b - E_a)/kT} = \frac{N_b}{N_a}, \tag{3.30}$$

where N_a is the number of atoms with energy E_a and N_b is the number of atoms with energy E_b.

As an example, let us now consider calculating the temperature at which we can expect equal numbers of atoms in the ground state ($n = 1$) and in the first excited state ($n = 2$). This is of interest because the Lyman series of absorption lines arise from atoms that start in the first excited state. There are two ground states corresponding to the two spin states allowed for the electron, so $g_1 = 2$. In the first excited state, there are four orbitals corresponding to the s and p orbitals and two electron spin states allowed for each orbital, so there are a total of eight excited states, so $g_2 = 8$. The energy of the ground state is $E_1 = -13.6\,\text{eV} = -E_0$, and the energy of the first excited state is $E_2 = E_1/4$. Now, we require $N_1 = N_2$, and so

$$1 = \frac{N_2}{N_1} = \frac{8}{2} e^{-3E_0/4kT}. \tag{3.31}$$

This is solved by:

$$T = \frac{3E_0}{4k \ln 4} = 8.54 \times 10^4\,\text{K}. \tag{3.32}$$

Therefore, at about 85,000 K, there should be roughly equal numbers of atoms in the excited state as there are in the ground state. Observationally, the maximum intensity of hydrogen lines coming from absorption in the first excited state comes at about $T = 9500$ K, so we are missing something in the analysis. What we are missing is the fact that at higher temperatures, the atoms can also be ionized. Since an ionized hydrogen atom cannot absorb a photon, only the total number of neutral hydrogen atoms contribute to the strength of the absorption lines. As the temperature increases, the total number of absorbers decreases even as the relative number of atoms in the excited state increases. Therefore in order to accurately describe the strength of the hydrogen lines in stellar spectra we also need to calculate the number of atoms in the ionized state relative to the number of atoms in the un-ionized state.

3.4 Boltzmann and Saha Equations

The Saha equation provides a way of calculating the number of atoms in different ionized states. A neutral atom is in ionization state $i = I$, a singly ionized atom is in the $i = II$ state, the doubly ionized atom is in the $i = III$ state, and so on. We define ξ_i as the ionizing energy required to remove an electron from an atom in the ground state taking it from ionization state i to ionization state $i+1$. The partition function gives the weighted sum of the number of ways an atom can have a given energy:

$$Z = \sum_{j=1}^{\infty} g_j e^{-(E_j - E_1)/kT}, \qquad (3.33)$$

where E_1 is the ground state energy for the given ionization. The ratio of the number of atoms in two adjacent ionization states is

$$\frac{N_{i+1}}{N_i} = \frac{2Z_{i+1}}{n_e Z_i} \left(\frac{2\pi m_e kT}{h^2} \right)^{3/2} e^{-\xi_i/kT}. \qquad (3.34)$$

For an element X, this equation results from considering the thermodynamic equilibrium of the reaction $X_{i+1} + e \rightleftharpoons X_i$. We can also express this in terms of the *free electron pressure* $P_e = n_e kT$ (where we have assumed an ideal electron gas):

$$\frac{N_{i+1}}{N_i} = \frac{2kT Z_{i+1}}{P_e Z_i} \left(\frac{2\pi m_e kT}{h^2} \right)^{3/2} e^{-\xi_i/kT}. \qquad (3.35)$$

Now, the strength of the Lyman lines in a star are proportional to the ratio of the number of HI atoms in the second excited state (N_2) to the total number of hydrogen atoms (N_{total}). Note that N_{total} includes both HI and HII atoms. Since we know from observation that the temperature of interest is around 10^4 K, we can assume that all the HI atoms are in either the ground or first excited state; therefore the total number of HI atoms is $N_I \sim N_1 + N_2$. Thus,

$$\frac{N_2}{N_{\text{total}}} = \frac{N_2}{N_I} \frac{N_I}{N_{\text{total}}} = \left(\frac{N_2}{N_1 + N_2} \right) \left(\frac{N_I}{N_I + N_{II}} \right)$$

$$= \left(\frac{N_2/N_1}{1 + N_2/N_1} \right) \left(\frac{1}{1 + N_{II}/N_I} \right). \qquad (3.36)$$

Now, for hydrogen, we have $N_2/N_1 = 4 e^{-\beta \Delta E}$, where $\beta = 1/kT$ and $\Delta E = E_2 - E_1 = \frac{3}{4} E_0$. There are no electron energy levels for an ionized hydrogen atom, so $Z_{II} = 1$, and the leading terms of the sum for Z_I are

$$Z_I = 2 + 8 e^{-\beta \Delta E} + \ldots, \qquad (3.37)$$

so

$$\frac{N_{II}}{N_I} = \frac{(m_e kT/2\pi \hbar^2)^{3/2}}{n_e (1 + 4 e^{-\beta \Delta E})} e^{-\beta \xi_1} \qquad (3.38)$$

Fig. 3.10 Fraction of hydrogen atoms that are in the first excited state for a stellar atmosphere with $P_e = 20\,\text{N}/\text{m}^2$. The rise at low temperatures is due to the increased number of excited atoms, while the drop at higher temperature is due to the increased number of completely ionized atoms

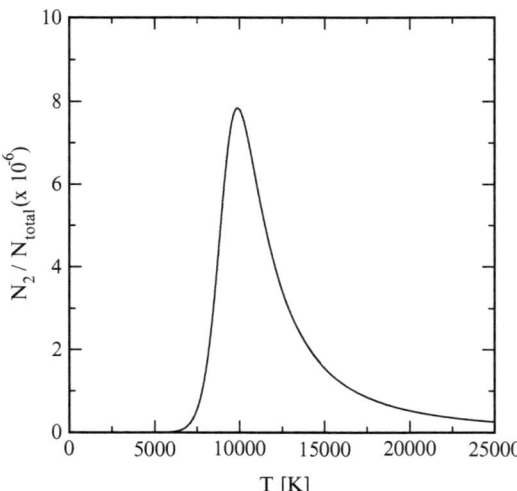

and finally,

$$\frac{N_2}{N_{\text{total}}} = \left(\frac{4e^{-\beta\Delta E}}{1+4e^{-\beta\Delta E}}\right)\left(1+\frac{(m_e kT/2\pi\hbar^2)^{3/2}}{n_e(1+4e^{-\beta\Delta E})}e^{-\beta\xi_1}\right)^{-1}. \quad (3.39)$$

This equation can be seen to peak at about $T = 9900\,\text{K}$ as shown in Fig. 3.10.

Problem 3.2: For a gas of neutral hydrogen atoms, at what temperature is the number of atoms in the first excited state only 1% of the number of atoms in the ground state? At what temperature is the number of atoms in the first excited state 10% of the number of atoms in the ground state?

Problem 3.3: A typical atmosphere found on a white dwarf of spectral type DB is pure helium. The ionization energies of neutral helium and singly ionized helium are $\xi_I = 24.6\,\text{eV}$ and $\xi_{II} = 54.4\,\text{eV}$, respectively. The partition functions are $Z_I = 1$, $Z_{II} = 2$, and $Z_{III} = 1$. Use $P_e = 20\,\text{N}/\text{m}^2$ for the electron pressure.

(a) Use the Saha equation to find N_{II}/N_I and N_{III}/N_{II} for temperatures of 5000 K, 15000 K, and 25000 K.
(b) Show that $N_{II}/N_{\text{total}} = N_{II}/(N_I + N_{II} + N_{III})$ can be expressed in terms of the ratios N_{II}/N_I and N_{III}/N_{II}.
(c) Plot N_{II}/N_{total} for temperatures between 5000 K and 25000 K. What is the temperature for which $N_{II}/N_{\text{total}} = 0.5$?

Thus, through observations of binaries, we can find the masses and radii of stars. Through observations of spectra we can find the surface temperatures and compositions of stars. Through observations of parallax we can find the total luminosities of stars. Plotting the luminosity vs. the mass, we find the mass–luminosity relation. Plotting the luminosity vs. the temperature, we find the H-R diagram which indicates that most stars lie along the main sequence. We want to understand and model all these relations. In Part II, we look at the physics needed to build models of stars.

Problems

3.1. GK Vir is an eclipsing spectroscopic binary with an angle of inclination $i = 89.5° \pm 0.6°$, so that it can be considered to be viewed edge on. The orbit is circular with semi-amplitudes of the radial velocities given by $K_1 = 38.6$ km/s and $K_2 = 221.6$ km/s. The orbital period is $P = 0.344$ d. The time required for the light curve to drop to its lowest value is $t_b - t_a = 89.6$ s, while the time required for the light curve to begin rising again is $t_c - t_a = 817$ s. Use this information to find:

(a) The radii of both stars
(b) The masses of both stars

3.2. For a gas of neutral hydrogen atoms, at what temperature is the number of atoms in the first excited state only 1% of the number of atoms in the ground state? At what temperature is the number of atoms in the first excited state 10% of the number of atoms in the ground state?

3.3. A typical atmosphere found on a white dwarf of spectral type DB is pure helium. The ionization energies of neutral helium and singly ionized helium are $\xi_I = 24.6$ eV and $\xi_{II} = 54.4$ eV, respectively. The partition functions are $Z_I = 1$, $Z_{II} = 2$, and $Z_{III} = 1$. Use $P_e = 20$ N/m² for the electron pressure.

(a) Use the Saha equation to find N_{II}/N_I and N_{III}/N_{II} for temperatures of 5,000 K, 15,000 K, and 25,000 K.
(b) Show that $N_{II}/N_{\text{total}} = N_{II}/(N_I + N_{II} + N_{III})$ can be expressed in terms of the ratios N_{II}/N_I and N_{III}/N_{II}.
(c) Plot N_{II}/N_{total} for temperatures between 5,000 K and 25,000 K. What is the temperature for which $N_{II}/N_{\text{total}} = 0.5$?

Part II
Equations and Processes

The physics involved in the structure and evolution of stars covers a wide range of disciplines. Hydrodynamics and thermodynamics contribute at the macroscopic scale, while quantum mechanics and nuclear physics operate at the microscopic scale. Here we develop the equations governing stellar structure and evolution. We then explore in some detail the underlying physics as it pertains to the descriptive variables within these equations.

Chapter 4
Stellar Evolution Equations

We want to develop models of stellar evolution that can reproduce the mass–luminosity relation as well as the structure of the H-R diagram. Obviously, we will start with several simplifying assumptions that can be relaxed as we try to achieve better fidelity to observations. These assumptions are:

1. Spherical symmetry
2. Isolation
3. Uniform initial composition

With spherical symmetry, we can describe the physical properties of stars as functions of r alone. One of these properties is the mass enclosed within a radius r:

$$m(r) = \int_0^r 4\pi r^2 \rho(r) dr, \qquad (4.1)$$

$$dm = \rho 4\pi r^2 dr. \qquad (4.2)$$

Problem 4.1: Using the density distribution

$$\frac{M}{4R^2} \frac{\sin(\pi r/R)}{r},$$

compute $m(r)$.

It is frequently more advantageous to use m instead of r as the independent variable since m is bounded in the range $0 \leq m \leq M$, where M is the total mass of the star. During the course of stellar evolution, the total mass may decrease slightly due to stellar winds, but there is not a wide fluctuation. On the other hand, the radius of the star will vary by several orders of magnitude during this time while there is minimal change in mass.

We also assume that the star is in *local thermodynamic equilibrium*. This allows us to calculate all of the thermodynamic properties in terms of the temperature T. Technically, thermodynamic equilibrium implies that all interactions and processes happen at the same rate as the inverse processes. "Local" thermodynamic equilibrium is achieved when the mean free path of particles in the gas is much smaller than the length scale for temperature change. The composition of a star is described by the mass fractions, X_i, of each element in the star.

With these assumptions in hand, we are ready to develop the equations governing stellar evolution. These equations will eventually allow for the structure and evolution of a star to be described in terms of three functions, $\rho(m)$, $T(m)$, and $X_i(m)$.

4.1 The Energy Equation

The energy equation for stellar structure is obtained by considering a small mass element dm that is in a spherical shell over which the temperature T, density ρ, and composition X_i can be considered constant. We let the internal energy per unit mass in this shell be u and define P to be the pressure and V to be the volume. According to the first law of thermodynamics, any changes in the internal energy are related to the heat added and the work done through the equation

$$\delta(u\,dm) = dm\,\delta u = \delta Q + \delta W. \tag{4.3}$$

The work done is

$$\delta W = -P\delta V = -P\delta\left(\frac{dV}{dm}dm\right) = -P\delta\left(\frac{1}{\rho}\right)dm. \tag{4.4}$$

We have used $dm/dV = \rho$ and we note that the thickness of the shell is defined in terms of dm, so that it can be considered a constant, although dr may vary as the internal energy and work change.

The heat added can come from the release of energy in the gas (usually from nuclear burning) or from an imbalance between the heat added and the heat removed due to flux through the stellar envelope. We will cover stellar envelopes and atmospheres in Chap. 6 and nuclear burning in Chap. 7. For the time being, let us define q to be the nuclear energy release per unit mass and $F(m)$ to be the heat flowing through the spherical surface defined by m, as shown in Fig. 4.1. At the surface of the star, $m = M$ and $F(M) = L$. Consequently the heat added is the change in these quantities over a time interval δt,

$$\delta Q = q\,dm\,\delta t + F(m)\delta t - F(m+dm)\delta t. \tag{4.5}$$

4.1 The Energy Equation

Fig. 4.1 Thin spherical shell of thickness dm enclosing a mass of m in a star with total mass M. The heat flow into the shell is $F(m)$ and the heat flow out of the shell is $F(m+dm)$

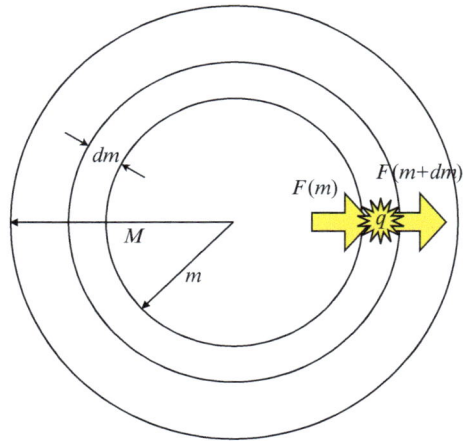

From the definition of the derivative, we have

$$F(m+dm) = F(m) + \frac{\partial F}{\partial m}dm, \quad (4.6)$$

so

$$\delta Q = \left(q - \frac{\partial F}{\partial m}\right) dm \delta t. \quad (4.7)$$

Finally, we have

$$dm \delta u = dm \left(q - \frac{\partial F}{\partial m}\right) \delta t - P \delta \left(\frac{1}{\rho}\right) dm. \quad (4.8)$$

Dividing by δt and converting the quantities $\delta/\delta t$ into time derivatives give the energy equation:

$$\dot{u} - \frac{P}{\rho^2}\dot{\rho} = q - \frac{\partial F}{\partial m}. \quad (4.9)$$

Note that this equation is valid even when the star is evolving. If the star is thermally stationary (often referred to as thermal equilibrium), we assume that the quantities do not vary in time, so $\dot{u} = 0$ and $\dot{\rho} = 0$. In this case the energy equation becomes

$$q = \frac{dF}{dm}, \quad (4.10)$$

and the nuclear energy generation rate is

$$L_{\text{nuc}} = \int_0^M q \, dm = \int_0^M dF = L. \quad (4.11)$$

As would be expected, we see that the luminosity of a star in equilibrium is equal to the nuclear energy released in its interior.

4.2 Hydrodynamic Equation

Consider a small volume element in a star given by $dV = drdS$ where dS is a surface area element at radius r, as shown in Fig. 4.2. The mass in this volume is $\Delta m = \rho dr dS$ (note that it is not dm because we are not looking at the mass of a spherical shell). The forces acting on this mass element are

Gravitation: $\quad -Gm\Delta m/r^2$
Pressure: $\quad P(r)dS - P(r+dr)dS$.

P is not a function of θ or ϕ because we have imposed spherical symmetry. Again, from the definition of the derivative, we have

$$P(r+dr) = P(r) + \frac{\partial P}{\partial r}dr, \qquad (4.12)$$

so

$$\Delta m \ddot{r} = -\frac{Gm\Delta m}{r^2} - \frac{\partial P}{\partial r}drdS,$$

$$= -\frac{Gm\Delta m}{r^2} - \frac{\partial P}{\partial r}\frac{\Delta m}{\rho}. \qquad (4.13)$$

As a consequence of gravitationally driven systems, the mass of the mass element can be divided out, leaving

$$\ddot{r} = -\frac{Gm}{r^2} - \frac{\partial P}{\partial r}\frac{1}{\rho}. \qquad (4.14)$$

Since we prefer to write everything in terms of m rather than r, we note that $dr = dm/4\pi r^2 \rho$ and so

$$\ddot{r} = -\frac{Gm}{r^2} - 4\pi r^2 \frac{\partial P}{\partial m}. \qquad (4.15)$$

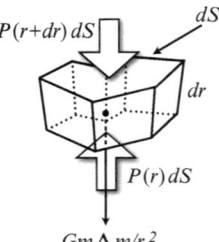

Fig. 4.2 The forces acting on a volume element of mass Δm in the interior of a star

This equation describes time-dependent systems that may be out of equilibrium and expanding or contracting. In equilibrium, $\ddot{r} = 0$, and we find the equation for hydrostatic equilibrium:

$$\frac{dP}{dm} = -\frac{Gm}{4\pi r^4}. \tag{4.16}$$

With the energy equation and the hydrodynamic equation, we can learn a number of interesting properties about the relationship between the temperature and size of a star in different situations. These are discussed later in this chapter.

4.3 Composition Equations

As we shall see later in more detail, stars shine by nuclear fusion. This means that the composition of the star changes with time. This can have a profound impact on the structure of the star as the nuclear fuel is used up or as the internal pressure changes. For example, consider the pressure and temperature near the center of the star made up of an ideal gas. The ideal gas law can be written

$$P = \frac{\rho}{m_g} kT, \tag{4.17}$$

where k is the Boltzmann constant. If the initial composition is hydrogen and after time all of the hydrogen has been converted to helium, the mass of the gas particles (m_g) has quadrupled. In order to maintain the pressure needed to support the outer layers of the star, the density or the temperature (or both) must increase. Thus, the evolution of the stellar composition contributes to the evolution of the structure of a star.

We describe the composition of the star in terms of the mass fraction of each nuclear species (e.g., H, He), defined as

$$X_i = \frac{\rho_i}{\rho}, \tag{4.18}$$

where ρ_i and ρ are the bulk densities in some region of the star. Frequently, when describing the bulk composition of a star, we use X for the mass fraction of hydrogen and Y for the mass fraction of helium, and then we lump all the other elements into a category called *metals*, described by the mass fraction Z. The number density of a given species is then

$$n_i = \frac{\rho_i}{m_i}, \tag{4.19}$$

where m_i is the mass of one nucleus of the given species. To a good approximation, we can write the mass of one nucleus as the sum of the masses of its nucleons:

$$m_i \simeq \mathscr{A}_i m_\mathrm{H}, \tag{4.20}$$

where $\mathscr{A}_i$ is the baryon number and m_H is 1/12 of the mass of a ^{12}C nucleus. Thus, we have

$$n_i = \frac{\rho_i}{\mathscr{A}_i m_H}, \tag{4.21}$$

and so we can relate X_i and n_i by the following:

$$n_i = \frac{\rho}{m_H} \frac{X_i}{\mathscr{A}_i}, \tag{4.22}$$

$$X_i = n_i \frac{\mathscr{A}_i}{\rho} m_H. \tag{4.23}$$

Nuclear processes inside stars result in changes in n_i. At the same time, the rate of nuclear reactions depends upon n_i as well as the temperature. In general, a nuclear reaction consists of two nuclei combining to create two different nuclei. Let us define $I(\mathscr{A}_i, \mathscr{Z}_i)$ and $J(\mathscr{A}_j, \mathscr{Z}_j)$ to be the reactants with baryon number $\mathscr{A}$ and nuclear charge $\mathscr{Z}$. The products of the reaction are given by $K(\mathscr{A}_k, \mathscr{Z}_k)$ and $L(\mathscr{A}_l, \mathscr{Z}_l)$. The reaction is then described by

$$I(\mathscr{A}_i, \mathscr{Z}_i) + J(\mathscr{A}_j, \mathscr{Z}_j) \rightleftharpoons K(\mathscr{A}_k, \mathscr{Z}_k) + L(\mathscr{A}_l, \mathscr{Z}_l) \tag{4.24}$$

subject to baryon conservation ($\mathscr{A}_i + \mathscr{A}_j = \mathscr{A}_k + \mathscr{A}_l$) and charge conservation ($\mathscr{Z}_i + \mathscr{Z}_j = \mathscr{Z}_k + \mathscr{Z}_l$). Several reactions will also include neutrons ($\mathscr{A} = 1$, $\mathscr{Z} = 0$), electrons ($\mathscr{A} = 0$, $\mathscr{Z} = -1$), and positrons ($\mathscr{A} = 0$, $\mathscr{Z} = +1$). Protons and neutrons are part of a class of particles called baryons. Positrons and electrons are part of a class of particles called leptons. Leptons are also conserved, so whenever a positron or electron is produced in a reaction, additional leptons known as neutrinos must also be produced in order to conserve lepton number. Note that the reaction can go both ways.

In order to obtain an equation relating the time evolution of the composition of stars, we need to look at the rates of nuclear reactions. From terrestrial experiments using a beam of nuclei fired at a target, we can measure an effective cross section for a reaction using

$$\sigma(E) = \frac{\text{number of reactions/nucleus/time}}{\text{number of incident particles/area/time}}. \tag{4.25}$$

It has units of area and can be thought of as roughly the cross-sectional area of the target particle to incoming particles with energy E. To find the reaction rates in units of reactions/volume/time, we need to consider the number of particles that will hit a target of cross-sectional area $\sigma(E)$, assuming that all the particles are moving in one direction. Let x denote a target particle and i denote an incident particle. If the gas is described by Maxwell-Boltzmann statistics, then

$$n_E dE = \frac{2n}{\sqrt{\pi}} \frac{1}{(kT)^{3/2}} E^{1/2} e^{-E/kT} dE, \tag{4.26}$$

4.3 Composition Equations

Fig. 4.3 Cylinder of volume $\sigma(E)v(E)\mathrm{d}t$ containing particles that will be incident on the target in time $\mathrm{d}t$

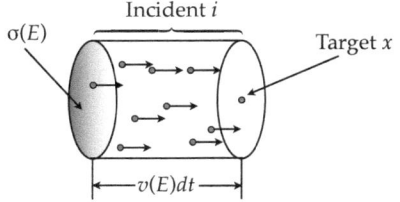

where $n_E \mathrm{d}E$ is the number density of particles with energies between E and $E + \mathrm{d}E$. Then the number of reactions ($\mathrm{d}N_E$) is the number of particles with energy E that can strike x in a time interval $\mathrm{d}t$ with a velocity $v(E) = \sqrt{2E/m_i}$. This is the number of particles contained in a volume $\sigma(E)v(E)\mathrm{d}t$, so:

$$\mathrm{d}N_E = \sigma(E)v(E)n_{iE}\mathrm{d}E\mathrm{d}t, \tag{4.27}$$

as shown in Fig. 4.3. Here, $n_{iE}\mathrm{d}E$ is the number density of incident particles with energies between E and $E + \mathrm{d}E$, which is a fraction of the total number of particles ($n_i = \int_0^\infty n_{iE}\mathrm{d}E$), so

$$n_{iE}\mathrm{d}E = \frac{n_i}{n}n_E\mathrm{d}E, \tag{4.28}$$

where n and n_E refer to all particles in the gas and n_i and n_{iE} refer only to species i. The number of reactions per target nucleus per time interval $\mathrm{d}t$ having energies between E and $E + \mathrm{d}E$ is

$$\frac{\text{reactions/nucleus}}{\text{time}} = \frac{\mathrm{d}N_E}{\mathrm{d}t} = \sigma(E)v(E)\frac{n_i}{n}n_E\mathrm{d}E. \tag{4.29}$$

Since there are n_x target particles/volume, the reaction rate is then

$$r_{ix} = \int_0^\infty n_x n_i \sigma(E) v(E) \frac{n_E}{n} \mathrm{d}E. \tag{4.30}$$

Note that if x is of the same type of particle as i then we have counted each interaction twice (once with the particle as target and once with the particle as incident), so the reaction rate for identical particles is

$$r_{ii} = \frac{1}{2} \int_0^\infty n_x n_i \sigma(E) v(E) \frac{n_E}{n} \mathrm{d}E. \tag{4.31}$$

We can simplify this equation by defining

$$R_{ijk} = \int_0^\infty \sigma(E) v(E) \frac{n_E}{n} \mathrm{d}E \sim \zeta v, \tag{4.32}$$

where ζ and v are averaged cross sections and velocities and the assumption is that R_{ijk} governs the reactions $I+J \to K+L$ (i.e., L is implied by I, J, and K and the conservation laws). Consequently, one can write the rate of change of n_i as

$$\dot{n}_i = -n_i \sum_{j,k} n_j R_{ijk} + \sum_{k,l} \frac{n_l n_k}{(1+\delta_{lk})} R_{lki}, \qquad (4.33)$$

where the first term describes the depletion of species i through the forward reaction and the second term describes the increase in species i due to the reverse reaction. Since the mass fraction is related to the number fraction, we have

$$\dot{X}_i = \frac{\mathscr{A}_i \rho}{m_H} \left(-\frac{X_i}{\mathscr{A}_i} \sum_{j,k} \frac{X_j}{\mathscr{A}_j} R_{ijk} + \sum_{l,k} \frac{X_l X_k}{\mathscr{A}_l \mathscr{A}_k} \frac{R_{lki}}{1+\delta_{lk}} \right). \qquad (4.34)$$

We can write this as a vector with each component representing a different species:

$$\dot{\mathbf{X}} = \mathbf{f}(\rho, T, \mathbf{X}), \qquad (4.35)$$

where $\mathbf{X} = (X_1, X_2, \ldots)$. Although Eq. (4.35) may seem unsatisfyingly vague, it is the equation governing the evolution of the composition of the star. It is usually implemented in stellar evolution using a table of known values of R_{ijk} for each reaction.

We now have the set of evolution equations describing the dynamics of the internal structure of a star:

$$\ddot{r} = -\frac{Gm}{r^2} - 4\pi r^2 \frac{\partial P}{\partial m}, \qquad (4.36)$$

$$\dot{u} - P\left(\frac{1}{\rho^2}\right)\dot{\rho} = q - \frac{\partial F}{\partial m}, \qquad (4.37)$$

$$\dot{\mathbf{X}} = \mathbf{f}(\rho, T, \mathbf{X}). \qquad (4.38)$$

This set includes the unknown structure functions $\rho(m,t)$, $T(m,t)$, and $\mathbf{X}(m,t)$ that must be solved in order to describe the star, but it also contains other functions that must be supplied from additional physics. Thermodynamics and statistical physics will give P and u, atomic physics and radiation transfer will supply F, and nuclear and particle physics will provide q and $\mathbf{f}$. We will look at these processes in more detail the next chapters.

4.4 Virial Theorem

We can use the equation of hydrostatic equilibrium to relate the gravitational energy of a star to its internal energy. We multiply Eq. (4.16) by the volume $V = \frac{4}{3}\pi r^3$ to obtain

$$V\frac{dP}{dm} = -\frac{1}{3}\frac{Gm}{r} \tag{4.39}$$

and note that

$$V dP = d(PV) - P dV = d(PV) - \frac{P}{\rho}dm$$

$$= -\frac{1}{3}\frac{Gm\,dm}{r}. \tag{4.40}$$

If we integrate over the entire star, we find that

$$\int_{\text{center}}^{\text{surface}} d(PV) = 0, \tag{4.41}$$

because $V = 0$ at the center and $P(M) = 0$ at the surface. Therefore, we have

$$-3\int_0^M \frac{P}{\rho}dm = -\int_0^M \frac{Gm\,dm}{r} = \Omega, \tag{4.42}$$

where Ω is the total gravitational potential energy. Thus,

$$-3\int_0^M \frac{P}{\rho}dm = \Omega. \tag{4.43}$$

This equation relating the pressure and density of a star to its gravitational potential energy is the general, global form of the *virial theorem*. We can apply this to the particular case of an ideal gas of particles with mass m_g to see a relationship between the temperature and the total internal energy and the gravitational potential energy of a star.

We expect the gas particles in stellar interiors to be completely ionized atoms, and so we can assume that the gas is monatomic. Assuming it is an ideal gas, we have

$$P = \left(\frac{\rho}{m_g}\right)kT. \tag{4.44}$$

In this case, the internal energy of the gas is simply the kinetic energy of the gas particles; therefore,

$$u = \frac{3}{2}\frac{kT}{m_g} = \frac{3}{2}\frac{P}{\rho}. \tag{4.45}$$

Therefore,
$$-3\int_0^M \frac{P}{\rho}dm = \Omega = -2\int_0^M u\,dm = -2U, \qquad (4.46)$$

and so $U = -\frac{1}{2}\Omega$. Note also that

$$U = \int_0^M \frac{3}{2}\frac{kT}{m_g}dm = \frac{3}{2}\frac{k}{m_g}\int_0^M T\,dm = \frac{3}{2}\frac{k\bar{T}M}{m_g}, \qquad (4.47)$$

where the average temperature is defined to be

$$\bar{T} = \frac{1}{M}\int_0^M T\,dm. \qquad (4.48)$$

Therefore the average temperature of a star is

$$\bar{T} = \frac{2m_g U}{3kM} = -\frac{1}{3}\frac{m_g \Omega}{kM}. \qquad (4.49)$$

From dimensional arguments, the gravitational energy of the star can always be written as $\Omega = -\alpha GM^2/R$ where α is some constant that depends upon the mass distribution of the star. For reasonable mass distributions, $\alpha \leq 1$. Thus,

$$\bar{T} = \frac{\alpha}{3}\frac{m_g G}{k}\frac{M}{R}. \qquad (4.50)$$

Writing R in terms of the average density $\bar{\rho} = 3M/4\pi R^3$ shows that

$$\bar{T} = \frac{\alpha}{3}\frac{m_g G}{k}\left(\frac{4\pi}{3}\right)^{1/3} M^{2/3}\bar{\rho}^{1/3} \qquad (4.51)$$

so that for stars of equal mass, the denser star is the hotter star. This implies that if a star contracts, it heats up and if it expands, it cools.

Problem 4.2: Using the density from Problem 4.1, compute Ω and show that $\alpha = 0.75$.

4.5 Total Energy

In order to determine the influences on the total energy of the star, we return to the energy equation and integrate over the entire star to find

$$\int_0^M \dot{u}\,dm + \int_0^M P\frac{d}{dt}\left(\frac{1}{\rho}\right)dm = L_{\text{nuc}} - L \qquad (4.52)$$

4.5 Total Energy

and exchange the time derivative with the mass integration to find

$$\int_0^M \dot{u}\,dm = \frac{d}{dt}\int_0^M u\,dm = \frac{d}{dt}U = \dot{U} \tag{4.53}$$

and

$$\int_0^M P\frac{d}{dt}\left(\frac{1}{\rho}\right)dm = \int_0^M P\left(\frac{\partial}{\partial t}\left(\frac{\partial V}{\partial m}\right)\right)dm$$

$$= \int_0^M P\frac{\partial \dot{V}}{\partial m}dm$$

$$= \int_0^M \frac{\partial}{\partial m}(P\dot{V})\,dm - \int_0^M \dot{V}\frac{\partial P}{\partial m}dm$$

$$= -\int_0^M 4\pi r^2 \dot{r}\frac{\partial P}{\partial m}dm, \tag{4.54}$$

so finally, we have

$$\dot{U} - \int_0^M 4\pi r^2 \dot{r}\frac{\partial P}{\partial m}dm = L_{\text{nuc}} - L. \tag{4.55}$$

This equation rates the total internal energy of a star to its nuclear generation rate and the luminosity. One integral remains to be evaluated so that this equation depends only on global quantities of the star. In order to evaluate this integral, we return to the equation of motion and integrate it over the star after multiplying by $\dot{r}$:

$$\ddot{r} = -\frac{Gm}{r^2} - 4\pi r^2 \frac{\partial P}{\partial m},$$

$$\dot{r}\ddot{r} = -\frac{Gm}{r^2}\dot{r} - 4\pi r^2 \dot{r}\frac{\partial P}{\partial m},$$

$$\int_0^M \dot{r}\ddot{r}\,dm = -\int_0^M \frac{Gm}{r^2}\dot{r}\,dm - \int_0^M 4\pi r^2 \dot{r}\frac{\partial P}{\partial m}dm. \tag{4.56}$$

The first integral can be evaluated to obtain

$$\int_0^M \dot{r}\ddot{r}\,dm = \int_0^M \frac{d}{dt}\left(\frac{1}{2}\dot{r}^2\right)dm = \frac{d}{dt}\int_0^M \frac{1}{2}\dot{r}^2\,dm = \dot{K}, \tag{4.57}$$

where K is the radial kinetic energy, which is also the total kinetic energy since we have assumed spherical symmetry. The second integral can also be evaluated, giving

$$-\int_0^M \frac{Gm}{r^2}\dot{r}\,dm = \int_0^M \frac{d}{dt}\left(\frac{1}{r}\right)Gm\,dm = \frac{d}{dt}\int_0^M \frac{Gm\,dm}{r} = -\dot{\Omega}. \tag{4.58}$$

Finally, we have

$$-\int_0^M 4\pi r^2 \dot{r} \frac{\partial P}{\partial m} dm = \dot{K} + \dot{\Omega}, \qquad (4.59)$$

and so we find an equation relating the total energies of the star:

$$\dot{U} + \dot{K} + \dot{\Omega} = L_{\text{nuc}} - L. \qquad (4.60)$$

Since $L_{\text{nuc}} - L$ is a measure of the net loss or gain of energy in the star, we can define the total energy of the star to be $E = U + K + \Omega$. And so $\dot{E} = L_{\text{nuc}} - L$.

We can see that if $L \neq L_{\text{nuc}}$, then the total energy must change. If the star is in thermal equilibrium so that $L_{\text{nuc}} = L$, then $\dot{E} = 0$. If it is also in hydrostatic equilibrium, then $\dot{K} = 0$. By the virial theorem, $U \propto -\Omega$. Therefore, if a star is in both thermal and hydrostatic equilibrium, $\dot{U} = 0$ and $\dot{\Omega} = 0$, it cannot expand and cool, nor can it contract and heat up.

Because $U \propto -\Omega$, we can set $U = -\alpha\Omega$, where α is a constant of proportionality. If the star is in hydrostatic equilibrium but not thermal equilibrium, then $E = U + \Omega$, and so

$$E = -\alpha\Omega + \Omega = (1-\alpha)\Omega = -\frac{(1-\alpha)}{\alpha}U. \qquad (4.61)$$

Note that if $\alpha < 1$, then $(1-\alpha)/\alpha > 0$, and so when $\dot{E} < 0$, then $\dot{U} > 0$ and therefore the temperature increases. All reasonable stellar models satisfy $\alpha < 1$. Thus, removing energy from the star will increase its temperature. Therefore, the star has a negative heat capacity.

4.6 Timescales

Once these additional functions have been expressed in terms of the unknowns (ρ, T, X_i), then we must determine boundary conditions. The two space (or mass) derivatives in Eqs. (4.36) and (4.37) require two spatial boundary conditions. The three time derivatives in Eqs. (4.36) and (4.37) plus the n time derivatives for the n species in Eq. (4.38) require $n + 3$ initial time conditions. The spatial boundary conditions are simple. At the surface of the star, the pressure must be zero; therefore $P(M, t) = 0$. At the center of the star, there cannot be a singularity of energy, so $F(0, t) = 0$. If we knew the initial state of the star, the initial conditions could come from defining $\dot{r}(m, 0)$, $\rho(m, 0)$, $T(m, 0)$, and $X_i(m, 0)$. Unfortunately, we do not know all of these functions. We can find a way out of this problem by looking at the timescales of the physical processes governed by these equations. The evolution equations each involve different types of change. Each of these changes has a typical timescale which we define as

$$\tau = \frac{\Theta}{\dot{\Theta}}, \qquad (4.62)$$

4.6 Timescales

where Θ is some physical quantity that changes due to a physical process.

The changes in the size of the star due to dynamical processes are described by Eq. (4.36). The dynamical timescale can therefore be set by looking at changes in R. Under the influence of gravity, the typical rate of change of R would be the escape velocity $v_{\rm esc} \sim \dot{R} = \sqrt{2GM/R}$, so

$$\tau_{\rm dyn} \sim \frac{R}{\dot{R}} = \sqrt{\frac{R^3}{2GM}}. \qquad (4.63)$$

Expressing this in terms of the average density and ignoring factors of order unity, we have $\bar{\rho} \sim M/R^3$, and so

$$\tau_{\rm dyn} \sim \frac{1}{\sqrt{G\bar{\rho}}}. \qquad (4.64)$$

We compare this with the only star for which we have intimate knowledge and use solar units to find

$$\tau_{\rm dyn} \sim (1000 \text{ s}) \sqrt{\left(\frac{R}{R_\odot}\right)^3 \left(\frac{M_\odot}{M}\right)} \qquad (4.65)$$

which is about 15 min This is a very short timescale compared to the lifetime of the sun, and it implies that imbalances between pressure and gravity are either (a) quickly restored to equilibrium or (b) quickly accelerated to catastrophic events. Since stars tend to live a lot longer than 15 min we can assume that stars are in a state of hydrostatic equilibrium throughout their lifetimes with their internal structure adjusting to maintain equilibrium. Small oscillations can occur about equilibrium, and these typically have timescales of a few minutes. If the internal structure cannot recover equilibrium quickly enough, we get a collapse or an explosion.

Changes in the internal energy of the star due to thermodynamic processes are governed by Eq. (4.37). The thermal timescale then can be set by changes in the internal energy U. Because the dynamical timescale is so short compared with known thermal variations of the sun, we can assume that the star is in hydrodynamic equilibrium and that the virial theorem holds, so that $U \sim GM^2/R$. The rate at which U changes is set by the rate at which energy is radiated away from the star ($\dot{U} = L$), so

$$\tau_{\rm th} \sim \frac{U}{L} \sim \frac{GM^2}{RL}. \qquad (4.66)$$

For typical solar units, we have

$$\tau_{\rm th} \sim \left(10^{15} \text{ s}\right) \left(\frac{M}{M_\odot}\right)^2 \left(\frac{R_\odot}{R}\right) \left(\frac{L_\odot}{L}\right), \qquad (4.67)$$

which is approximately 3×10^7 year. Although this is significantly longer than the dynamical timescale, it is still much smaller than the age of the sun. For example, the last major extinction of the dinosaurs occurred around 60 million years (or $2\tau_{\rm th}$) ago. The consequences of this timescale is that we can consider a star to be in thermodynamic equilibrium throughout most of its life. If the star also maintains hydrostatic equilibrium, we can assume that total energy is conserved and that (due to the virial theorem) gravitational potential energy and thermal energy are each separately conserved. This has profound consequences that we shall see later. If the star contracts quasi-statically in one part, then it must expand in another in order to conserve Ω. If there is a temperature increase in one part of the star, then it must be accompanied by a temperature decrease in another in order to conserve U. The thermal timescale can also be thought of as the time it would take a star to emit its entire reserve of thermal energy through gravitational contraction if the luminosity were held constant. The thermal timescale is sometimes referred to as the "Kelvin-Helmholtz" timescale.

The nuclear timescale changes the rest mass energy (as well as the species abundances) of the star. These effects are described by the set of equations in Eq. (4.38). We can relate the fraction ε of total rest mass energy that is released in typical nuclear reactions to the luminosity of the star L, so

$$\tau_{\text{nuc}} \sim \frac{\varepsilon M c^2}{L} = \varepsilon \left(4.5 \times 10^{20}\,\text{s}\right) \left(\frac{M}{M_\odot}\right) \left(\frac{L_\odot}{L}\right). \tag{4.68}$$

We estimate ε by considering the typical fraction of rest mass energy released by a nucleus compared to the nuclear rest mass energy. For helium this is $\varepsilon \sim 0.007$. For other nuclei, this is smaller, so we can approximate it by $\varepsilon \sim 10^{-3}$. Thus, $\tau_{\text{nuc}} \sim 10^{17}$ s for the sun. This is about ten times its age. Consequently, the sun has not burned all of its fuel and it is the rate of burning that governs the rate of stellar evolution.

Another consequence of the different timescales is that one can assume that the star is in thermodynamic equilibrium for changes that occur on timescales greater than τ_{th} and it is in hydrostatic equilibrium on timescales greater than τ_{dyn}. Therefore, we can work with a simplified set of evolution equations when looking at the long-term evolution of stars by assuming both hydrostatic and thermodynamic equilibrium:

$$\frac{dP}{dm} = -\frac{Gm}{4\pi r^4}, \tag{4.69}$$

$$\frac{dF}{dm} = q, \tag{4.70}$$

$$\dot{\mathbf{X}} = \mathbf{f}(\rho, T, \mathbf{X}). \tag{4.71}$$

Problem 4.3: Assuming that a star of mass M is devoid of nuclear energy sources, determine its radius as a function of time, if it maintains a constant luminosity L. Assume that the star is in hydrostatic equilibrium.

Problems

4.1. Using the density distribution

$$\frac{M}{4R^2} \frac{\sin(\pi r/R)}{r},$$

compute $m(r)$.

4.2. Using the density from Problem 4.1, compute Ω and show that $\alpha = 0.75$.

4.3. Assuming that a star of mass M is devoid of nuclear energy sources, determine its radius as a function of time, if it maintains a constant luminosity L. Assume that the star is in hydrostatic equilibrium.

Chapter 5
Gas and Radiation Pressures

In order to solve the evolution equations [Eqs. (4.36)–(4.38)] for the variables ρ, T, and X, we need to be able to express the auxiliary functions P, F, and q in terms of these variables. Additional physics must be included to accomplish this. In this chapter, we concentrate on the pressure. Equations that relate the pressure to the density, temperature, and composition are known as *equations of state*. The most familiar equation of state is the ideal gas law, which we have already seen:

$$PV = NkT \longrightarrow P = \frac{N}{V}kT = \frac{\rho}{m}kT. \tag{5.1}$$

The ideal gas law assumes that the constituent gas particles are noninteracting point particles that obey classical (i.e., non-quantum) statistics. In order to determine when the ideal gas law can be used as the equation of state for stellar interiors, we need to estimate the degree to which the gas particles can be considered noninteracting.

At the expected temperatures in the interior of stars, the gas will consist of completely ionized atoms. Therefore the dominant interaction will be via the Coulomb force. The interaction can be characterized by the electrostatic energy between constituent gas particles. If this energy is significantly smaller than the average kinetic energy due to thermal motion, then any interaction will only alter the thermal energy of the particle in a negligible way. In this case, the particles can be considered to be essentially noninteracting.

The average kinetic energy due to thermal motion is comparable to $k\bar{T}$, where $\bar{T}$ is the average temperature of the gas throughout the star. Using $\bar{T}$ as calculated in Eq. (4.50), we have

$$k\bar{T} \simeq \frac{\alpha}{3}\frac{GM}{R}\mathscr{A}m_\text{H}, \tag{5.2}$$

where $\mathscr{A}$ is the average baryon number for the gas particles and α is the proportionality constant introduced in Chap. 4. The typical electrostatic potential energy between two particles is

$$\varepsilon_C \simeq \frac{1}{4\pi\varepsilon_0}\frac{\mathscr{Z}^2 e^2}{d}, \qquad (5.3)$$

where $\mathscr{Z}e$ is the average charge of the gas particles. We estimate the separation distance d by assuming that the average separation is the length of the side of a cube containing a single particle; thus,

$$d = \left(\frac{\mathscr{A} m_H}{\bar{\rho}}\right)^{1/3} = \left(\frac{4\pi \mathscr{A} m_H}{3M}\right)^{1/3} R, \qquad (5.4)$$

where M and R are the total mass and radius of the star, respectively. The ratio of Coulomb energy to the thermal kinetic energy is then

$$\frac{\varepsilon_C}{k\bar{T}} = \left\{\frac{3}{\alpha^{3/4}\varepsilon_0 4\pi}\right\}^{4/3} \frac{\mathscr{Z}^2 e^2}{G(\mathscr{A} m_H)^{4/3} M^{2/3}}. \qquad (5.5)$$

Assuming that $\alpha \sim 1$, we find the factor $\{3/\alpha^{3/4} 4\pi\}^{4/3} \sim 0.1$. Further assuming that the bulk composition of the star is hydrogen (so that $\mathscr{A} = \mathscr{Z} = 1$), then,

$$\frac{\varepsilon_C}{k\bar{T}} \sim 0.01 \left(\frac{M_\odot}{M}\right)^{2/3}. \qquad (5.6)$$

Consequently, the Coulomb interaction contributes less than 1% to the typical energies involved in particle interactions and so we can use the ideal gas law.

Another assumption made in using the ideal gas law is that the particles interact classically as particles rather than quantum mechanically as waves. In order to estimate the degree to which quantum effects play a role in the interactions between gas particles, we will look at the typical de Broglie wavelength of a gas particle compared with the average particle separation. If the de Broglie wavelength is small compared with the separation, then there is negligible interference of the particle wavefunctions, and so quantum effects can be ignored. Assuming nonrelativistic gas particles, we have

$$\lambda = \frac{h}{p} = \frac{h}{\sqrt{2mE}} = \frac{h}{\sqrt{2\mathscr{A} m_H k\bar{T}}}, \qquad (5.7)$$

and so

$$\frac{\lambda}{d} = \left(\frac{3}{2G}\right)^{1/2}\left(\frac{3}{4\pi}\right)^{1/3}\frac{h}{(\mathscr{A} m_H)^{4/3}}\left(\frac{1}{M^{1/6} R^{1/2}}\right)$$

$$= 0.01 \left(\frac{M_\odot}{M}\right)^{1/6}\left(\frac{R_\odot}{R}\right)^{1/2}. \qquad (5.8)$$

Therefore, for typical solar values, the interactions are dominated by classical effects. Thus, we can conclude that the ideal gas law is valid for the interiors of typical stars.

5.1 Gas Pressure

A gas that consists of several different species of particles can be described by an ideal gas law for each species. The total pressure in the star is then the sum of the pressures due to each species. We usually break the pressures up into three types: P_I = ion pressure, P_e = electron pressure, and $P_{\rm rad}$ = radiation pressure. The total pressure is then

$$P = P_I + P_e + P_{\rm rad} = P_{\rm gas} + P_{\rm rad}. \tag{5.9}$$

We can define a parameter β that gives the fraction of the total pressure due to gas, so that $P_{\rm gas} = \beta P$ and $P_{\rm rad} = (1-\beta)P$.

Gas pressure arises from many different ions, and the number of electrons will depend upon the temperature and the types of ions. We will start by looking at the ion pressure, which is a sum over the pressures due to each ion species:

$$P_I = \sum_i P_i = \sum_i n_i kT = \sum_i \left(\frac{\rho}{m_{\rm H}} \frac{X_i}{\mathscr{A}_i}\right) kT, \tag{5.10}$$

where we used Eq. (4.23) in the last step. Remembering that X_i is the mass fraction of each species, we have

$$\sum_i \left(\frac{X_i}{m_{\rm H}\mathscr{A}}\right) = \sum_i \frac{n_i}{\rho} = \frac{\sum n_i}{\rho} = \frac{n}{\rho} = \frac{1}{\bar{m}}, \tag{5.11}$$

where $\bar{m}$ is the average mass of a particle of the gas. We define the μ to be the *mean molecular weight* or *mean atomic mass* to be the "atomic mass" of an average mass so $\bar{m} = \mu_I m_{\rm H}$, and

$$\frac{1}{\mu_I} = \sum_i \frac{X_i}{\mathscr{A}_i}. \tag{5.12}$$

Since the majority of the matter in a main sequence star is made up of hydrogen and helium, we separate them out and lump all other elements in a group called metals. As in Chap. 4, we define X = mass fraction of hydrogen, Y = mass fraction of helium, and Z = mass fraction of metals. We find

$$\frac{1}{\mu_I} = X + \frac{Y}{4} + \frac{Z}{\langle\mathscr{A}\rangle} = X + \frac{Y}{4} + \frac{1-X-Y}{\langle\mathscr{A}\rangle}, \tag{5.13}$$

where $\langle \mathscr{A} \rangle$ is the average atomic number of all the metals, so it depends upon the number and type of metals in the star. For metal composition typical of the sun, $\langle \mathscr{A} \rangle \sim 15.5$. Defining $\mathscr{R} = k/m_H$, we have for the ion pressure

$$P_I = \frac{\mathscr{R}}{\mu_I}\rho T. \tag{5.14}$$

Now we can turn our attention to the electron pressure, given by

$$P_e = n_e kT. \tag{5.15}$$

In the interiors of stars, we can safely assume that every atom is completely ionized, so the electron number density is

$$n_e = \sum_i \mathscr{Z}_i n_i = \frac{\rho}{m_H}\sum_i \frac{X_i \mathscr{Z}_i}{\mathscr{A}_i}. \tag{5.16}$$

We define

$$\frac{1}{\mu_e} = \sum_i \frac{X_i \mathscr{Z}_i}{\mathscr{A}_i}, \tag{5.17}$$

which is the average number of electrons per nucleon in the gas (so μ_e is the average number of nucleons per electron). This leads to $n_e = \rho/\mu_e m_H$, and

$$\frac{1}{\mu_e} = X + \frac{Y}{2} + (1-X-Y)\left\langle \frac{\mathscr{Z}}{\mathscr{A}} \right\rangle, \tag{5.18}$$

where $\langle \mathscr{Z}/\mathscr{A} \rangle$ is the average number of electrons per nucleon for the metals. This can be reasonably approximated by 1/2, so

$$\frac{1}{\mu_e} = X + \frac{1}{2}Y + \frac{1}{2} - \frac{1}{2}X - \frac{1}{2}Y = \frac{1}{2}(1+X). \tag{5.19}$$

Note that as the star burns hydrogen, the value of μ_e increases. The electron pressure is

$$P_e = \frac{\mathscr{R}}{\mu_e}\rho T, \tag{5.20}$$

and so the total gas pressure is

$$P_{\text{gas}} = P_I + P_e = \left(\frac{1}{\mu_I} + \frac{1}{\mu_e}\right)\mathscr{R}\rho T = \frac{\mathscr{R}}{\mu}\rho T, \tag{5.21}$$

where the total mean molecular weight of the gas is defined through

$$\frac{1}{\mu} = \frac{1}{\mu_I} + \frac{1}{\mu_e}. \tag{5.22}$$

5.2 Radiation Pressure

The ideal gas law comes from assuming that the pressure arises from the transfer of momentum from the particles in a gas to the walls of an imaginary container. The calculation of the force exerted on a hypothetical surface due to a system of particles with momentum distribution $n(p)dp$ and velocities v results in an integral known as the pressure integral:

$$P = \frac{1}{3}\int_0^\infty v p n(p) dp. \tag{5.23}$$

The pressure integral defines the pressure in terms of the momentum flux through a unit surface area. For nonrelativistic massive particles, we use the Maxwell–Boltzmann distribution of momentum:

$$n(p)dp = \frac{n 4\pi p^2 dp}{(2\pi m k T)^{3/2}} e^{-p^2/2mkT} \tag{5.24}$$

and so obtain the ideal gas law.

Problem 5.1: Evaluate the pressure integral using the Maxwell–Boltzmann distribution and show that you get the ideal gas law.

The pressure integral is equally valid for photons, which also carry momentum. In this case, the momentum is expressed in terms of the light frequency by $p = h\nu/c$. Consequently, using the Planck distribution of frequencies for a blackbody:

$$n(\nu)d\nu = \frac{8\pi \nu^2}{c^3} \frac{d\nu}{\left(e^{h\nu/kT} - 1\right)}, \tag{5.25}$$

the pressure integral gives

$$P = \frac{1}{3}\int_0^\infty c\left(\frac{h\nu}{c}\right) \frac{8\pi \nu^2}{c^3} \frac{d\nu}{\left(e^{h\nu/kT} - 1\right)}. \tag{5.26}$$

The radiation pressure is found from integrating Eq. (5.26):

$$P_{\text{rad}} = \left(\frac{8\pi^5 k^4}{15 c^3 h^3}\right)\frac{1}{3}T^4 = \frac{4}{3}\frac{\sigma}{c}T^4, \tag{5.27}$$

where σ is the Stefan–Boltzmann constant. The combination $4\sigma/c$ is sometimes called the *radiation constant a*, so $P_{\text{rad}} = \frac{1}{3}aT^4$.

5.3 Degeneracy Pressure

If the wavefunctions of the gas particles begin to overlap, we will need to take into account quantum effects such as the uncertainty principle and the exclusion principle. This can occur if the density of the gas is sufficiently high. If the temperature does not increase enough, then a significant fraction of the lowest-energy states can become filled and collisional energy cannot elevate the low-energy state particles up to the energy of unfilled states. We will consider here only electrons, since their behavior is important during the evolution of stars. Nucleons become important only at the very end of the evolution of massive stars, and the extension of this formalism to nucleons is quite straightforward.

We begin by determining the energy levels of electrons confined to a region of space defined by the volume of the star. Consider a cubic box of volume L^3; then the wave functions for free particles confined to the box have wavelengths:

$$\lambda_x = \frac{2L}{N_x}, \quad \lambda_y = \frac{2L}{N_y}, \quad \lambda_z = \frac{2L}{N_z}; \quad (5.28)$$

consequently, the momentum is $p = h/\lambda$. If the gas is not relativistic, then $E = p^2/2m$, so

$$E = \frac{1}{2m}\left(p_x^2 + p_y^2 + p_z^2\right) = \frac{h^2}{8mL^2}\left(N_x^2 + N_y^2 + N_z^2\right) = \frac{h^2 N^2}{8mL^2}, \quad (5.29)$$

where $N^2 = N_x^2 + N_y^2 + N_z^2$. Fundamentally, this is a result of the uncertainty principle in that the minimum momentum of a particle is constrained by the maximum volume that the particle may occupy.

From the exclusion principle, we know that no two fermions can occupy exactly the same state. If the temperature is 0 K, then all states up to some maximum value

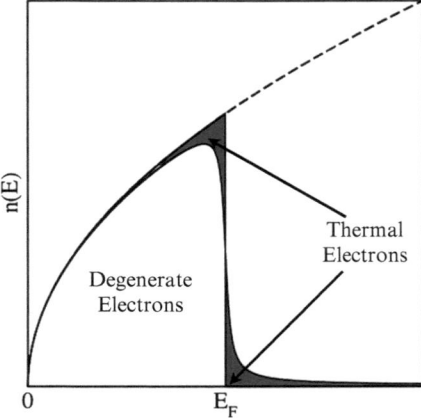

Fig. 5.1 Number of occupied states as a function of energy. The *vertical line* indicates the number of filled states at $T = 0$; the *shaded area* indicates the number of electrons that can contribute to the thermal energy of the gas

5.3 Degeneracy Pressure

of N will be filled with electrons. Since there are two spin states allowed for each energy state (N_x, N_y, N_z), the total number of electrons in the gas (N_e) that are in the ground state of the entire population of electrons in the star corresponds to twice the total number of unique quantum numbers N_x, N_y, and N_z. Since there are a very large number of electrons in a typical stellar core, the discreteness of the states washes out and the number of unique quantum numbers corresponds to the volume of one-eighth of a sphere of radius $N = \sqrt{N_x^2 + N_y^2 + N_z^2}$, so

$$N_e = 2 \left(\frac{1}{8}\right)\left(\frac{4}{3}\pi N^3\right). \tag{5.30}$$

Solving for N gives

$$N = \left(\frac{3N_e}{\pi}\right)^{1/3}. \tag{5.31}$$

Thus, at $T = 0$, all the electron states with energies below a certain energy will be filled with electrons. This is known as the *Fermi energy* and is found by substituting Eq. (5.31) into Eq. (5.29) to obtain

$$E_F = \frac{\hbar^2}{2m}\left(3\pi^2 n_e\right)^{2/3}, \tag{5.32}$$

where $n_e = N_e/L^3$ is the number density of electrons. Since all of the states below this energy are filled, the number of electrons with energy between E and $E + dE$ is equal to twice the number of states in this range, so

$$n(E)dE = 2\left(\frac{1}{8}\right)4\pi N^2 dN = \frac{\pi}{2h^3}(8m)^{3/2} E^{1/2} dE, \tag{5.33}$$

where we have used

$$N = \sqrt{\frac{8mL^2 E}{h^2}}. \tag{5.34}$$

This condition holds when the temperature of the gas is 0 K. If the temperature is above 0, then some electrons will have energies above the Fermi energy (Fig. 5.1).

The amount of extra energy is about $\frac{3}{2}kT$. Therefore, electrons with energies above $E_F - \frac{3}{2}kT$ can move to higher energy states through collisions, while those with energies less than this amount have no empty state to move up to. We can relate the condition for degeneracy to a limit on the temperature and electron number density of the gas. In particular, we note that the gas is completely nondegenerate if $E_F - \frac{3}{2}kT \leq 0$ because all electrons can exchange energy and therefore contribute to the ideal gas equation of state. Thus, the condition for degeneracy is $E_F \geq \frac{3}{2}kT$. In order to relate this condition to the primary quantities ρ, T, and **X**, we note that the number of electrons per unit volume can be calculated by

$$n_e = \left(\frac{\mathscr{Z}}{\mathscr{A}}\right)\frac{\rho}{m_H}, \tag{5.35}$$

so

$$E_F = \frac{\hbar^2}{2m_e}\left[3\pi^2\left(\frac{\mathscr{Z}}{\mathscr{A}}\right)\frac{\rho}{m_H}\right]^{2/3}. \tag{5.36}$$

Setting $\frac{3}{2}kT \leq E_F$ for degeneracy, we have

$$\frac{3}{2}kT < \frac{\hbar^2}{2m_e}\left[3\pi^2\left(\frac{\mathscr{Z}}{\mathscr{A}}\right)\frac{\rho}{m_H}\right]^{2/3} \tag{5.37}$$

or

$$\frac{T}{\rho^{2/3}} < \frac{\hbar^2}{3km_e}\left[3\pi^2\left(\frac{\mathscr{Z}}{\mathscr{A}}\right)\frac{1}{m_H}\right]^{2/3}. \tag{5.38}$$

For typical elements above hydrogen, $(\mathscr{Z}/\mathscr{A}) = 0.5$, and this condition reads

$$\frac{T}{\rho^{2/3}} < 1261 \text{ K m}^2/\text{kg}^{2/3}. \tag{5.39}$$

The smaller that $T/\rho^{2/3}$ is, the more degenerate the gas.

The equation of state for a degenerate gas is very different than that for an ideal gas. In particular, the pressure is independent of the temperature. To see this, we calculate the average energy of a degenerate electron:

$$\bar{E} = \frac{1}{N_e}\int_0^{E_F} g(E)E\,dE, \tag{5.40}$$

where $g(E) = dN_s/dE$ is the density of states where

$$N_s = \frac{\pi}{3}\left(\frac{E}{E_0}\right)^{3/2}, \tag{5.41}$$

with

$$E_0 = \frac{h^2}{8m_e L^2}. \tag{5.42}$$

We find that

$$\bar{E} = \frac{3}{5}E_F, \tag{5.43}$$

and so the pressure is

$$P = -\frac{\partial E}{\partial V} = -\frac{\partial}{\partial V}\left(\frac{3}{5}N_e E_F\right) = \frac{1}{5}\frac{\hbar^2}{m_e}(3\pi^2)^{2/3}n_e^{5/3}$$

$$= \frac{1}{5}\frac{\hbar^2(3\pi^2)^{2/3}}{m_e}\left[\frac{\mathscr{Z}}{\mathscr{A}}\frac{\rho}{m_H}\right]^{5/3} = K\rho^{5/3}. \tag{5.44}$$

5.4 Internal Energy of Gas and Radiation

Note that this calculation is based on the assumption that the electrons were non-relativistic (i.e., $E = p^2/2m$).

Problem 5.2: The relativistic form of the kinetic energy is $K = (\gamma - 1)mc^2$. Determine the density for which $\gamma \sim 1.1$ (i.e., $p^2/2m \sim 0.1\,mc^2$).

In the event that the electrons start to become relativistic, we must change our energy calculations since the energy is then given by $E = \gamma mc^2$. The transition zone from nonrelativistic to relativistic regimes is quite difficult, but the extreme relativistic case can easily be obtained from assuming $v \simeq c$ so $E \simeq pc$. The result of this is that the equation of state is of the form

$$P = K'\rho^{4/3}. \tag{5.45}$$

Problem 5.3: For highly relativistic electrons, $E \simeq pc$ where $p = \sqrt{p_x^2 + p_y^2 + p_z^2}$. Following Eq. (5.28) we have $E = \hbar c \frac{\pi}{L}\sqrt{N_x^2 + N_y^2 + N_z^2} = \hbar c \pi N/L$. Derive Eq. (5.45) as follows:

(a) The number of states is related to N by $N_s = \left(\frac{\pi}{3}N^3\right)$. Use this to write E as a function of N_s and show that

$$g(E) = \frac{dN_s}{dE} = \frac{1}{\pi^2}\left(\frac{L}{\hbar c}\right)^3 E^2.$$

(b) Define $E_F = E(N_e)$ where N_e is the total number of free electrons in the star and show that

$$\bar{E} = \frac{1}{N_e}\int_0^{E_F} g(E)E\,dE = \frac{3}{4}E_F.$$

(c) Noting that $L = V^{1/3}$, show that

$$P = -\frac{\partial}{\partial V}(N_e\bar{E}) = \frac{1}{4}\hbar c\left(3\pi^2\right)^{1/3} n_e^{4/3}$$

where $n_e = N_e/V$.

5.4 Internal Energy of Gas and Radiation

The specific energy ($u =$ energy/mass) of a gas is given by

$$u = \frac{1}{\rho}\int_0^\infty n(p)E(p)\,dp, \tag{5.46}$$

where $E(p)$ is the kinetic energy in terms of the momentum and $n(p)$ is the number density of particles with momentum p. For a classical system, $E = p^2/2m$, but for a relativistic system,

$$E = mc^2 \left[\left(1 + \frac{p^2}{m^2c^2}\right)^{1/2} - 1 \right]. \tag{5.47}$$

If the gas is degenerate, we integrate up to the *Fermi momentum* p_F, which is defined such that $E_F = E(p_F)$. (We make the assumption that the gas is completely degenerate, so $T = 0$ and the partition function is a step function.) For a nonrelativistic gas, we have

$$u_{\text{gas}} = \frac{3}{2}nkT = \frac{3}{2}\frac{P_{\text{gas}}}{\rho}. \tag{5.48}$$

If we look at a degenerate gas and perform the integral, we obtain precisely the same result:

$$u_{\text{gas}} = \frac{3}{2}\frac{P_{\text{gas}}}{\rho}. \tag{5.49}$$

Problem 5.4: Do the integral in Eq. (5.46) for a classical degenerate gas to obtain Eq. (5.49).

If the degenerate gas is relativistic, then

$$u_{\text{gas}} = \frac{3P_{\text{gas}}}{\rho}. \tag{5.50}$$

For radiation, we replace the integral by

$$u_{\text{rad}} = \frac{1}{\rho}\int_0^\infty h\nu n(\nu)d\nu = \frac{aT^4}{\rho}. \tag{5.51}$$

Note that this is the energy density. Thus,

$$u_{\text{rad}} = \frac{3P_{\text{rad}}}{\rho}. \tag{5.52}$$

The similarity between radiation and relativistic degenerate gases is not coincidental. The relativistic condition is that $E \gg mc^2$, so that $E \simeq pc$, which is identical to the energy/momentum relation for photons.

5.5 Adiabatic Exponent

Processes that are rapid compared to the thermal timescale happen too quickly for there to be any heat exchange relative to the environment. These are called adiabatic processes. For adiabatic processes, the thermodynamic equation [Eq. (4.37)] reads

$$du - \frac{P}{\rho}\frac{d\rho}{\rho} = 0. \tag{5.53}$$

Note that for ideal gases and degenerate systems, $u \propto P/\rho$, so we can write

$$u = \phi \frac{P}{\rho}, \tag{5.54}$$

where ϕ is a proportionality constant. Thus,

$$du = \phi \frac{dP}{\rho} - \phi \frac{P}{\rho}\frac{d\rho}{\rho}, \tag{5.55}$$

and so

$$\phi \frac{dP}{\rho} - (1+\phi)\frac{P}{\rho}\frac{d\rho}{\rho} = 0. \tag{5.56}$$

This results in the following differential equation:

$$\frac{dP}{P} = \frac{1+\phi}{\phi}\frac{d\rho}{\rho} \tag{5.57}$$

whose solution is

$$P = K_a \rho^{\gamma_a} \tag{5.58}$$

where K_a is a constant of the integration that depends upon the entropy, and the adiabatic exponent is $\gamma_a = (\phi+1)/\phi$. Note that for nonrelativistic ideal and degenerate gases $\gamma_a = 5/3$, while for radiation and relativistic degenerate gases, $\gamma_a = 4/3$. Mixtures of radiation and partially relativistic degenerate gases will give $4/3 < \gamma_a < 5/3$.

It may seem that γ_a has a lower bound of $4/3$, but the discussion so far has centered on regions where the total number of particles is constant. This is valid in regions where the gases are completely ionized or completely neutral. However, there are intermediate regions where the ionization level is very dependent upon temperature and density. In these regions, we need to take into account the Saha equation as well. In general, this calculation can be quite complicated, so we will focus on a simple system of a nonrelativistic, ideal hydrogen gas that can only have two ionization states. We define the degree of ionization as

$$x = \frac{N_{II}}{N_{\text{total}}} = \frac{n_+}{n_0 + n_+}, \tag{5.59}$$

where n_+ is the number density of ions and n_0 is the number density of neutral atoms. The number density of electrons is determined by the number density of ions, so $n_e = n_+$. The Saha equation can be written as

$$\frac{n_+ n_e}{n_0} = \frac{g}{h^3}(2\pi m_e kT)^{3/2} e^{-\xi/kT}, \qquad (5.60)$$

where the partition function information is absorbed into the constant g.

We can write the gas pressure in terms of the number densities of ions, electrons, and atoms as

$$P_{\text{gas}} = P_0 + P_+ + P_e = (n_0 + n_+ + n_e)kT$$

$$= \left(\frac{n_0 + n_+ + n_+}{n_0 + n_+}\right)(n_0 + n_+)kT$$

$$= (1+x)(n_0 + n_+)kT = (1+x)(n_0 + n_+)\mathcal{R}\rho T. \qquad (5.61)$$

We engage in a little algebraic manipulation to write the left-hand side of the Saha equation as

$$\frac{n_+ n_e}{n_0} = \frac{n_+^2}{n_0} = \frac{\left[n_+^2/(n_0+n_+)^2\right](n_0 + 2n_+)}{\left[(n_0+n_+)^2 - n_+^2\right]/(n_0+n_+)^2}$$

$$= \frac{x^2}{1-x^2}\frac{P_{\text{gas}}}{kT} = \frac{g}{h^3}(2\pi m_e kT)^{3/2} e^{-\xi/kT}. \qquad (5.62)$$

The specific energy of a partially ionized gas also contains the potential energy of ionization. This is the energy that can be released when the ions recombine, so

$$u = \frac{3}{2}\frac{P}{\rho} + \xi\frac{n_+}{\rho}. \qquad (5.63)$$

The density can be written as $\rho = (n_0 + n_+)m_H$, and so

$$u = \frac{3}{2}\frac{P}{\rho} + \xi\frac{n_+}{(n_0+n_+)m_H} = \frac{3}{2}\frac{P}{\rho} + \frac{\xi}{m_H}x. \qquad (5.64)$$

Using Eq. (5.62), we can now compute du to find

$$du = \frac{3}{2}\frac{1}{\rho}dP - \frac{3}{2}\frac{P}{\rho}\frac{d\rho}{\rho} + \frac{\xi}{m_H}\left(\frac{\partial x}{\partial P}dP + \frac{\partial x}{\partial \rho}d\rho\right). \qquad (5.65)$$

Applying this result to the adiabatic condition and performing a hideous batch of algebra result in

$$\gamma_a = \frac{5 + \left(\frac{5}{2} + \frac{\xi}{kT}\right)^2 x(1-x)}{3 + \left(\frac{3}{2} + \left(\frac{3}{2} + \frac{\xi}{kT}\right)^2\right) x(1-x)}. \tag{5.66}$$

At either extreme of fully ionized or fully neutral gas, we find $\gamma_a = 5/3$. The minimum value occurs at $x = 0.5$ and depends upon T and ξ. In the limit that $\xi/kT \to \infty$, $\gamma_a \to 1$. The adiabatic exponent will become important as we look at the stability of stars against convection.

Problems

5.1. Evaluate the pressure integral using the Maxwell–Boltzmann distribution and show that you get the ideal gas law.

5.2. The relativistic form of the kinetic energy is $K = (\gamma - 1)mc^2$. Determine the density for which $\gamma \sim 1.1$. (i.e., $p^2/2m \sim 0.1\,mc^2$).

5.3. For highly relativistic electrons, $E \simeq pc$ where $p = \sqrt{p_x^2 + p_y^2 + p_z^2}$. Following Eq. (5.28) we have $E = \hbar c \frac{\pi}{L}\sqrt{N_x^2 + N_y^2 + N_z^2} = \hbar c \pi N/L$. Derive Eq. (5.45) as follows:

(a) The number of states is related to N by $N_s = \left(\frac{\pi}{3} N^3\right)$. Use this to write E as a function of N_s and show that

$$g(E) = \frac{dN_s}{dE} = \frac{1}{\pi^2}\left(\frac{L}{\hbar c}\right)^3 E^2.$$

(b) Define $E_F = E(N_e)$ where N_e is the total number of free electrons in the star and show that

$$\bar{E} = \frac{1}{N_e}\int_0^{E_F} g(E)E\,dE = \frac{3}{4}E_F. \tag{5.67}$$

(c) Noting that $L = V^{1/3}$, show that

$$P = -\frac{\partial}{\partial V}(N_e \bar{E}) = \frac{1}{4}\hbar c \left(3\pi^2\right)^{1/3} n_e^{4/3},$$

where $n_e = N_e/V$.

5.4. Do the integral in Eq. (5.46) for a classical degenerate gas to obtain Eq. (5.49).

Chapter 6
Radiative Transfer and Stellar Atmospheres

In order to obtain the expression for the transport of radiant energy through the layers of the star in terms of ρ, T, and $\mathbf{X}$, we need to spend a little time exploring the envelopes of stars. We will revisit the energy density and pressure due to radiation and discuss the consequences of local thermodynamic equilibrium. Finally, we will also look in more detail at the nature of absorption lines in stellar spectra.

6.1 The Radiation Field

In the energy equation, F is the heat flux through a surface of constant m in the star. This can come from conduction, convection, or radiation. We will discuss convection later in Chap. 9, while conduction will be discussed briefly in Chap. 11. Here we concentrate on the flow of heat through radiation. In order to talk about the transport of energy through radiation, we need to define a quantity that describes the radiation field and the flow of energy through that field. The heat flux is the total flow of heat energy through a surface, integrated over all directions and wavelengths. To obtain this value, we need to look at the energy carried in a specific direction at a specific wavelength. We define I_λ to be the "specific intensity" and it is the power per area emitted from a surface area element dA into a solid angle dΩ at an angle θ from the normal to dA within a wavelength range $(\lambda, \lambda + d\lambda)$. Although I_λ is not a vector, we draw it with an arrow in Fig. 6.1 to indicate the implied direction of energy flow.

The specific intensity may vary with direction, so in order to define a single quantity at a point in the interior of a star, we compute the "mean intensity" as the average of I_λ over all directions:

$$\langle I_\lambda \rangle = \frac{1}{4\pi} \int_0^{2\pi} \int_0^{\pi} I_\lambda \sin\theta \, d\theta \, d\phi. \tag{6.1}$$

Fig. 6.1 Specific intensity I_λ and its orientation relative to the solid angle dΩ and the emitting area dA

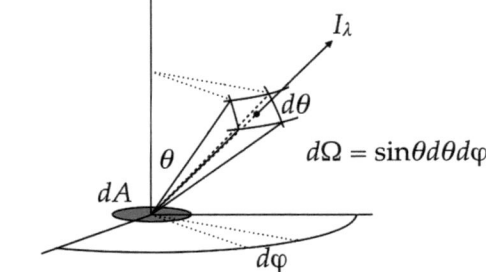

Fig. 6.2 Volume element $dV = dAdh$ and path of original radiation plus radiation entering from an adjacent volume element

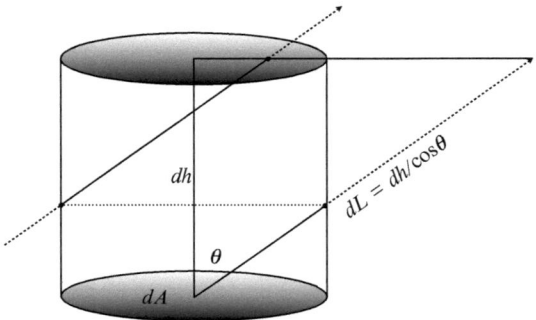

If the radiation field is entirely described by the blackbody spectrum, then the specific intensity is isotropic and $\langle I_\lambda \rangle = I_\lambda = B_\lambda$, where

$$B_\lambda = \frac{c}{4\pi}u(\lambda) = \frac{2hc^2}{\lambda^5 \left(e^{hc/\lambda kT} - 1\right)} \qquad (6.2)$$

is the specific intensity for blackbody radiation.

The energy carried by radiation in the wavelength range $(\lambda, \lambda + d\lambda)$ through dA in a time interval dt in the direction of dΩ is then

$$E_\lambda d\lambda = I_\lambda d\lambda dA \cos\theta d\Omega dt. \qquad (6.3)$$

We can use this expression to obtain the energy density of radiation for a radiation field given by an arbitrary I_λ. Consider a volume element that is projected normal to the emitting surface element dA so that $dV = dAdh$. Energy that is emitted in a given direction, θ, from the surface will travel through this volume until it passes through the side. However, radiation from an adjacent volume will enter from the opposite side to replace it. Consequently, we can set the time interval that this radiation spends inside the volume to be $dt = dL/c = dh/(c\cos\theta)$, as shown in Fig. 6.2. Therefore, the energy within the volume due to the radiation emitted at angle θ is

$$E_\lambda d\lambda = I_\lambda d\lambda dA d\Omega \cos\theta \frac{dh}{c\cos\theta} = \frac{I_\lambda d\lambda d\Omega}{c} dV. \qquad (6.4)$$

6.1 The Radiation Field

The total energy over all directions (including radiation coming from the top of the volume) is found by integrating over $d\Omega$, so

$$u(\lambda)d\lambda = \frac{1}{c}\int I_\lambda d\lambda d\Omega = \frac{4\pi}{c}\langle I_\lambda\rangle d\lambda. \tag{6.5}$$

We can recover the blackbody energy density [Eq. (3.19)] if we replace $\langle I_\lambda\rangle$ with B_λ.

From Chap. 5, we know that the pressure on a mathematical surface is related to the flow of momentum through that surface. We can understand this by imagining that the mathematical surface is replaced by a real, solid surface. The momentum that would normally flow through the surface must now bounce off of it, reversing direction in the component normal to the surface. This change in momentum amounts to a force on the surface. Thus, we can calculate the radiation pressure due to a radiation field given by I_λ by considering the flux of the component of momentum normal to a surface. If we take the surface element dA to lie in the xy-plane, then the pressure is

$$P_{\text{rad}} = \int \frac{dp_z}{dt\,dA} = \int \frac{dp\cos\theta}{dt\,dA}, \tag{6.6}$$

but $p = E/c$ for photons, and so for each wavelength interval, we have

$$dp_\lambda d\lambda = E_\lambda d\lambda/c = \frac{1}{c}I_\lambda d\lambda dA \cos\theta d\Omega dt \tag{6.7}$$

and

$$P_{\text{rad},\lambda}d\lambda = \frac{1}{c}\int I_\lambda d\lambda \cos^2\theta d\Omega = \frac{d\lambda}{c}\int I_\lambda \cos^2\theta d\Omega. \tag{6.8}$$

Therefore, the radiation pressure can be obtained from the radiation field using

$$P_{\text{rad}} = \frac{1}{c}\int_0^\infty d\lambda \int_0^{2\pi} d\phi \int_0^\pi d\theta I_\lambda \cos^2\theta \sin\theta. \tag{6.9}$$

Problem 6.1: Show that you can recover the expression for P_{rad} in Eq. (5.27) if you use B_λ, the blackbody expression for the specific intensity.

The energy flux, $\Phi_\lambda d\lambda$, passing through a surface area element dA is the net energy flow in a direction normal to the surface. Thus,

$$\Phi_\lambda d\lambda = \frac{E_\lambda d\lambda}{dA\,dt} = \int I_\lambda \cos\theta d\Omega. \tag{6.10}$$

If the radiation field is isotropic, then $\Phi_\lambda = 0$. Therefore, if we want a nonzero flux of energy through a surface, then I_λ must be a function of θ.

6.2 Radiative Transfer

As radiation passes through material in a star, energy can be added and removed from the beam due to interactions with matter. Energy is removed from the beam as photons are scattered into different directions and wavelengths or absorbed by atoms. Energy is added to the beam as photons are scattered into the direction of the beam or are emitted by atoms. These processes are sketched in Fig. 6.3. The equation giving the change in specific intensity as a function of distance traveled through a medium is known as the *transfer equation*. In a steady state, the change in I_λ along a path ds through a gas of density ρ is given by

$$dI_\lambda = -\kappa_\lambda \rho I_\lambda \, ds + j_\lambda \rho \, ds, \tag{6.11}$$

where κ_λ is the *opacity* of the gas at wavelength λ and j_λ is the *emission coefficient* of the gas. Dividing by $-\kappa_\lambda \rho$ yields the transfer equation:

$$-\frac{1}{\kappa_\lambda \rho} \frac{dI_\lambda}{ds} = I_\lambda - S_\lambda, \tag{6.12}$$

where the *source function*, $S_\lambda = j_\lambda / \kappa_\lambda$, is the ratio of emission to absorption coefficients and is a measure of how rapidly the photons in the beam are removed and replaced by photons in the local vicinity.

The opacity of the gas has dimensions of area/mass and arises from several different mechanisms. It is often strongly dependent on the wavelength of the photon and the temperature of the gas. The main processes that contribute to the opacity involve different interactions between photons and electrons, and they are described by the initial and final states of the electron. These are known as *bound–bound, bound–free, free–free*, and *electron scattering*.

Bound–Bound: This process involves the absorption of a photon, causing the transition of a bound electron to a higher bound energy level in an atom. Clearly,

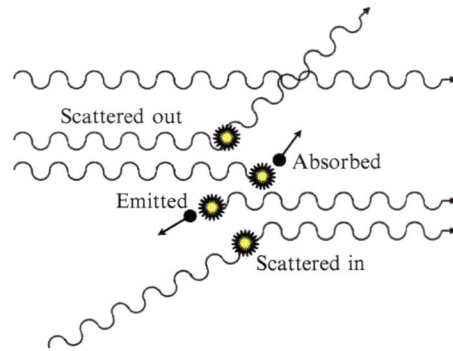

Fig. 6.3 Processes that add and remove energy from a beam as it passes through a gas. The scattering processes can both remove and add energy to the beam. Absorption removes energy, and emission adds energy

the strength of this absorption is highly wavelength dependent, but it also depends on the ionization and excitation states of the atoms in the gas. As we have seen in Sect. 3.4, the ionization and excitation levels are temperature dependent. Deep in the interiors of stars, the gas is completely ionized, and so bound–bound scattering is important only in the atmospheres. The reverse process when the electron drops to a lower energy level and emits a photon is included in the emission coefficient.

Bound–Free: This process occurs if a photon has sufficient energy to completely strip an electron from an atom and ionize it. In this case, the absorption will have a upper bound on the wavelength based on the minimum energy needed to ionize the atom, but all shorter wavelengths are allowed as the extra energy can be distributed as kinetic energy of the newly freed electron. As in the case of bound–bound absorption, the strength and wavelength dependence of this absorption vary according to the ionization and excitation levels of the atoms in the gas. Consequently, the opacity due to this process is temperature dependent and is important in the atmospheres. Recombination is an emission process.

Free–Free: Although a free electron cannot absorb a photon and simultaneously conserve energy and momentum, it is possible for a free electron to absorb a photon if it is in an unbound interaction with a nearby atom that can accept the excess energy and momentum. In this case, the electron is free before and after the absorption and so the opacity is nonzero across all wavelengths. The opposite of this effect is *bremsstrahlung*, when a decelerating electron emits a photon.

Electron Scattering: In the case of Thomson or Rayleigh scattering, a photon imparts some energy and momentum to an electron or atom and is deflected out of the beam and may also have its wavelength changed. Like free–free absorption, this process is also applicable across a continuum of wavelengths.

Problem 6.2: The relativistic momentum of a particle is $\gamma m v$ and the relativistic energy is $\gamma m c^2$, where $\gamma = \left(1 - v^2/c^2\right)^{-1/2}$. Use relativistic conservation of momentum and energy to show that a free particle cannot absorb a photon.

The quantity $\kappa_\lambda \rho$ has the dimensions of length^{-1} and can be thought of as the inverse of the mean free path of a photon of wavelength λ in a gas of density ρ. A dimensionless quantity called the *optical depth* is defined by

$$d\tau_\lambda = -\kappa_\lambda \rho ds \qquad (6.13)$$

and is a measure of the transparency of the gas. Usually the optical depth is measured backwards along the direction that the energy is being transported— hence the minus sign. It can also be thought of as giving the number of scattering or absorption events that a photon will have experienced during its passage along integrated path. We can then rewrite the transfer equation in terms of the optical depth to obtain

$$\frac{dI_\lambda}{d\tau_\lambda} = I_\lambda - S_\lambda. \qquad (6.14)$$

Fig. 6.4 Radial optical depth line element and its relationship to the true optical depth and the direction of the specific intensity beam for a small surface element inside of a star

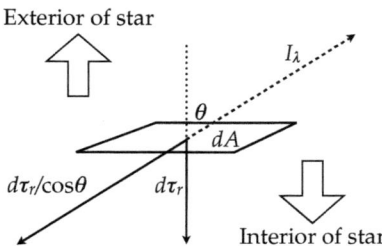

We are interested in finding the radiative energy flux through surfaces of constant r (or m). Therefore we want to solve the transfer equation for the specific intensity $I(\tau_r, \theta)$, where $\tau_r = \tau \cos\theta$ as shown in Fig. 6.4. We can do this through the use of an integrating factor applied to the differential equation:

$$\cos\theta \frac{dI_\lambda}{d\tau_{r\lambda}} = I_\lambda - S_\lambda. \tag{6.15}$$

Using

$$e^{-\tau_r/\cos\theta} \tag{6.16}$$

as the integrating factor and performing some simple algebra, we find

$$\frac{d}{d\tau_r}[I_\lambda(\tau_r,\theta)e^{-\tau_r/\cos\theta}] = -S_\lambda(\tau_r,\theta)\frac{e^{-\tau_r/\cos\theta}}{\cos\theta}. \tag{6.17}$$

This equation can be integrated to obtain the desired expression. The limits of integration will be from the surface of interest down to the center of the star. Technically, we need to know the optical depth at the center of the star, but we can take it to be infinity for all reasonable stellar models. Therefore, the solution gives

$$I_\lambda = \frac{1}{\cos\theta}\int_\infty^{\tau_r} S_\lambda(t,\theta)e^{(\tau_r-t)/\cos\theta} dt, \tag{6.18}$$

and the focus now is to find a sensible value of S_λ.

The interpretation of Eq. (6.18) is that the specific intensity at any given optical depth is a weighted sum of the source functions from all lower depths in the star. The weighting reflects the gradual extinction of the intensity from greater depths and the replacement of this intensity with the local intensities at each intervening layer. At large optical depths the weighting strongly favors nearby layers and the mean free path of the photons is short enough that the source function can be approximated by a Taylor expansion of the specific intensity for blackbody radiation

$$S_\lambda(t) = \sum_n \frac{(t-\tau)^n}{n!}\left.\frac{\partial^n B_\lambda}{\partial t^n}\right|_\tau = B_\lambda + (t-\tau)\left.\frac{\partial B_\lambda}{\partial t}\right|_\tau + \frac{(t-\tau)^2}{2}\left.\frac{\partial^2 B_\lambda}{\partial t^2}\right|_\tau + \cdots, \tag{6.19}$$

6.3 Radiative Heat Flux

and so the specific intensity is given by

$$I_\lambda = \int_\tau^\infty \sum_{n=0}^\infty \frac{(t-\tau)^n}{n!} \left.\frac{\partial^n B_\lambda}{\partial t^n}\right|_\tau e^{(t-\tau)/\cos\theta} \frac{dt}{\cos\theta}. \tag{6.20}$$

We can perform the integral by making a change of variables to $u = (t-\tau)/\cos\theta$ and noting that

$$\Gamma(n+1) = \int_0^\infty u^n e^{-u} du \tag{6.21}$$

to obtain

$$I_\lambda = \sum_{n=0}^\infty \cos^n\theta \frac{\partial^n B_\lambda}{\partial \tau^n}. \tag{6.22}$$

6.3 Radiative Heat Flux

We want to obtain an expression for the radiative heat flux using Eq. (6.22). At large optical depths, we only need to consider the leading nonzero term in the expansion. Therefore, using Eq. (6.10) we have

$$\Phi_\lambda = \int I_\lambda \cos\theta d\Omega = 2\pi \int_0^\pi \sum_{n=0}^\infty \cos^{n+1}\theta \frac{\partial^n B_\lambda}{\partial \tau^n} \sin\theta d\theta. \tag{6.23}$$

Only odd values of n survive, and the leading term is

$$\Phi_\lambda = \frac{4\pi}{3}\frac{\partial B}{\partial \tau} = \frac{4\pi}{3}\frac{\partial T}{\partial \tau}\frac{\partial B_\lambda}{\partial T}. \tag{6.24}$$

Expressed in terms of dr and the opacity, κ_λ, this is

$$\Phi_\lambda = -\frac{4\pi}{3\rho}\frac{\partial T}{\partial r}\frac{1}{\kappa_\lambda}\frac{\partial B_\lambda}{\partial T}. \tag{6.25}$$

From here we can get the heat flux using

$$F = 4\pi r^2 \int_0^\infty \Phi_\lambda d\lambda, \tag{6.26}$$

where the $4\pi r^2$ comes from integrating over the spherical surface of radius r. The precise value of F depends on a knowledge of the opacity as a function of λ, but we can sidestep this issue by introducing a quantity called the *Rosseland mean opacity* defined as

$$\bar{\kappa} = \frac{\int_0^\infty \frac{\partial B_\lambda}{\partial T} d\lambda}{\int_0^\infty \frac{1}{\kappa_\lambda}\frac{\partial B_\lambda}{\partial T} d\lambda}. \tag{6.27}$$

With the Rosseland mean opacity in hand, then the heat flux can be expressed as

$$F = -\frac{4\pi}{3\rho}\frac{\partial T}{\partial r}\frac{1}{\bar{\kappa}}\frac{\partial}{\partial T}\int_0^\infty B_\lambda d\lambda. \tag{6.28}$$

Fig. 6.5 Rosseland mean opacities calculated using the Opacity Project tools for a pure hydrogen gas ($X = 1$), a gas with no metals but a helium fraction of $Y = 0.2$, and a gas with a mixture of hydrogen, helium, and metals. The plots are calculated using $\log \rho / T_6^3 = -3$, where T_6 is the temperature in units of 10^6 K

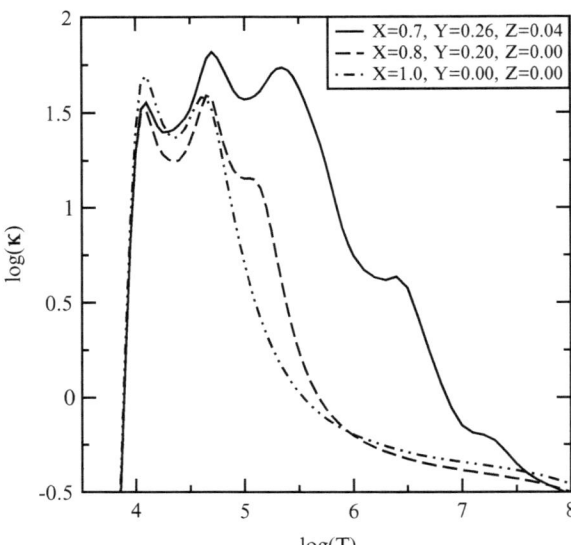

Recalling that $B_\lambda = u(\lambda) c / 4\pi$ and

$$\int_0^\infty u(\lambda) \mathrm{d}\lambda = aT^4 \tag{6.29}$$

gives the expression

$$F = 4\pi r^2 \frac{4acT^3}{3\bar{\kappa}\rho} \frac{\partial T}{\partial r}. \tag{6.30}$$

It remains to find $\bar{\kappa}$ as a function of ρ, T, and **X**. There exist online tools such as the Opacity Project that can produce $\bar{\kappa}$ through numerical computation, given an input concentration of several elements. A sample of the temperature dependence of the Rosseland mean opacity computed from the Opacity Project is shown in Fig. 6.5.

At lower temperatures, such as those found near the surface of a star, there may be a substantial fraction of bound electrons, and so the opacities due to bound–bound and bound–free interactions dominate. These are the most difficult opacities to compute since they arise from millions of absorption lines from all the different species in the gas. Consequently, at low temperatures, $\bar{\kappa}$ must be computed numerically and there are no simple asymptotic formulae.

At intermediate temperatures, such as those found in the upper envelopes of stars, the gas becomes substantially more ionized and bound–free and free–free processes dominate the opacities. The mean opacities for these processes are proportional to

$\rho T^{-7/2}$. Mean opacities that obey this relation are called *Kramers opacities*. The bound–free opacity roughly follows

$$\bar{\kappa}_{bf} \sim 4 \times 10^{21} Z(1+X)\rho T^{-7/2} \, \text{m}^2/\text{kg}, \tag{6.31}$$

and the free–free opacity is roughly two orders of magnitude lower, with

$$\bar{\kappa}_{ff} \sim 10^{19} \frac{Z^2}{\mu_e \mu_I} \rho T^{-7/2} \, \text{m}^2/\text{kg}. \tag{6.32}$$

At still higher temperature, such as those found deep inside stars, where the atoms are all completely ionized, then electron scattering is the dominant contributor to the opacity. As long as the average photon energy is well below the rest mass energy of the electron, then $\bar{\kappa}_{es}$ is simply the number of electrons per mass times the Thomson cross section of the electron:

$$\bar{\kappa}_{es} = \frac{n_e}{\rho} \sigma_e = \frac{n_e}{\rho} \frac{8\pi}{3} \left(\frac{e^2}{4\pi\varepsilon_0 m_e c^2} \right)^2 \simeq 0.02(1+X). \tag{6.33}$$

The total opacity is simply the sum of the individual opacities at a given temperature.

Now that we have the opacities in hand, we can express the radiative heat flux in terms of either r or m:

$$F = -\frac{(4\pi r^2) \, 4acT^3}{3\bar{\kappa}\rho} \frac{dT}{dr} = -\frac{(4\pi r^2)^2 \, 4acT^3}{3\bar{\kappa}} \frac{dT}{dm}. \tag{6.34}$$

6.4 Model Atmospheres

Before turning to the details of nuclear burning rates, we will use the tools developed in this chapter to look in more detail at the outer atmospheres of stars, where the approximation of large optical depth begins to break down. Generally, the atmosphere of a star is taken to be the region where the integrated optical depth from the surface (or $r = \infty$) is less than 1000. Most of the photons that leave a star come from this region. This is where we measure the spectral lines and the effective temperature.

In order to model the atmospheres, we make some simplifying assumptions based on the fact that the thickness of the atmosphere is very small compared to the radius of the star. Thus, we can ignore curvature effects and approximate the atmosphere as a flat, planar slab described by Cartesian coordinates with the xy-plane at the surface of the atmosphere and the z-axis extending out from the surface, as shown in Fig. 6.6. This picture is somewhat problematic, since the surface of a star is

Fig. 6.6 Coordinate description of plane-parallel model atmosphere, showing the radial optical depth as well as the true optical depth along a ray making an angle of θ with the vertical

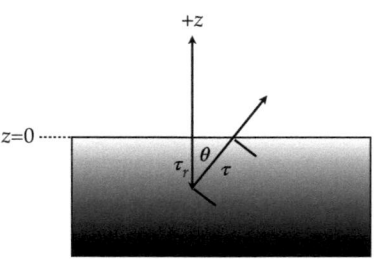

not an easily defined solid surface. However, we can use the radial optical depth as shown in Fig. 6.4, so that

$$\tau_{r\lambda} = \int_z^\infty \kappa_\lambda \rho \, dz \qquad (6.35)$$

and define the surface to be the point where the optical depth is small enough to be effectively zero. Returning to Eq. (6.15), we have

$$\cos\theta \frac{dI_\lambda}{d\tau_{r\lambda}} = I_\lambda - S_\lambda. \qquad (6.36)$$

Since the conditions near the surface of stars allow for bound–bound and bound–free interactions, computation of the opacities as a function of wavelength can be quite complicated. Fortunately, at wavelengths that are not associated with discrete transition lines, the opacity is dominated by electron scattering. As long as the photons involved do not have energies comparable to the rest mass energies of electrons, the electron scattering opacities are independent of wavelength. Consequently, if we ignore the discrete transition wavelengths, we can assume that the opacity is constant for all wavelengths. An atmosphere based on this assumption is called a *grey atmosphere*. For a grey atmosphere, we have

$$d\tau_{r\lambda} \longrightarrow d\tau_r \qquad (6.37)$$

and

$$\cos\theta \frac{dI}{d\tau_r} = I - S \qquad (6.38)$$

with

$$I = \int_0^\infty I_\lambda \, d\lambda, \qquad (6.39)$$

$$S = \int_0^\infty S_\lambda \, d\lambda. \qquad (6.40)$$

If we integrate Eq. (6.38) over all solid angles, we have

$$\frac{d}{d\tau_r} \int I \cos\theta \, d\Omega = \int I \, d\Omega - \int S \, d\Omega. \qquad (6.41)$$

6.4 Model Atmospheres

The source function is independent of direction, and using the definitions of the energy flux, Φ_λ, and the mean intensity, this becomes

$$\frac{d}{d\tau_r}\Phi = 4\pi(\langle I \rangle - S), \tag{6.42}$$

where $\Phi = \int \Phi_\lambda d\lambda$. We can also obtain an expression for the radiation pressure by integrating the first moment of Eq. (6.38):

$$\frac{d}{d\tau_r}\int I\cos^2\theta d\Omega = \int I\cos\theta d\Omega - \int S\cos\theta d\Omega. \tag{6.43}$$

Again, S is independent of direction, so the last term integrates to 0. The first term gives the derivative of cP_{rad}, and the second term is Φ. Therefore,

$$c\frac{dP_{\text{rad}}}{d\tau_r} = \Phi. \tag{6.44}$$

The equations above are valid for any grey atmosphere. Here we focus on a plane-parallel atmosphere in equilibrium. In this case, no heat is deposited in the atmosphere; therefore the flux through each layer is identical, and so

$$\frac{d\Phi}{d\tau_r} = 0 \Longrightarrow S = \langle I \rangle. \tag{6.45}$$

Also, the radiation pressure equation can now be integrated to give

$$P_{\text{rad}} = \frac{1}{c}\Phi\tau_r + \text{constant}. \tag{6.46}$$

From Eq. (5.27), we know that $P_{\text{rad}} = aT^4/3$, and so if we knew the value of the constant above, then we could determine the temperature as a function of depth in the atmosphere. Furthermore, we could then determine the depth in the atmosphere at which $T = T_{\text{eff}}$ and therefore know more about the environment in which the observed light from stars originates. One approach to determining the constant is to note that if the flux through each layer is identical, then the difference between outgoing and incoming intensity at each surface is a constant. We treat the outgoing intensity as being uniform over all angles $0 \leq \theta < \pi/2$, with value $\langle I_{\text{out}} \rangle$, and the incoming intensity as uniform over all angles $\pi/2 < \theta \leq \pi$ with value $\langle I_{\text{in}} \rangle$ At the very outside of the star, where there is no incoming intensity, $I = 0$ for $\theta \leq \pi/2$. Therefore, the outgoing intensity is related to the flux by

$$\Phi = \int_0^{\pi/2}\int_0^{2\pi} I\cos\theta\sin\theta d\theta d\phi \equiv \pi\langle I_{\text{out}} \rangle. \tag{6.47}$$

Knowing that Φ is constant allows us to determine the average ingoing intensity at any depth in terms of the average outgoing intensity at that depth. Therefore,

$$\Phi = \pi \left(\langle I_{\text{out}} \rangle - \langle I_{\text{in}} \rangle \right) \tag{6.48}$$

and the mean intensity is simply the average between the ingoing and outgoing intensities:

$$\langle I \rangle = \frac{1}{2} \left(\langle I_{\text{out}} \rangle + \langle I_{\text{in}} \rangle \right). \tag{6.49}$$

The radiation pressure can now be expressed in terms of $\langle I_{\text{out}} \rangle$ and $\langle I_{\text{in}} \rangle$ as

$$P_{\text{rad}} = \frac{2\pi}{3c} \left(\langle I_{\text{out}} \rangle + \langle I_{\text{in}} \rangle \right) = \frac{4\pi}{3c} \langle I \rangle. \tag{6.50}$$

Thus, at $\tau_r = 0$, we have

$$P_{\text{rad}} = \frac{4\pi}{3c} \langle I \rangle = \frac{2\pi}{3c} \langle I_{\text{out}} \rangle = \frac{2}{3c} \Phi = \text{constant}. \tag{6.51}$$

Now, knowing that the luminosity is related to Φ by

$$\Phi = \frac{L}{4\pi R^2} = \sigma T_{\text{eff}}^4, \tag{6.52}$$

we find

$$\langle I \rangle = \frac{3\sigma}{4\pi} T_{\text{eff}}^4 \left(\tau_r + \frac{2}{3} \right). \tag{6.53}$$

Finally, if we assume that the atmosphere is in local thermodynamic equilibrium, then $\langle I \rangle = \int B_\lambda d\lambda$, and so

$$\langle I \rangle = \frac{\sigma T^4}{\pi} = \frac{3\sigma}{4\pi} T_{\text{eff}}^4 \left(\tau_r + \frac{2}{3} \right). \tag{6.54}$$

This implies that T_{eff} is the temperature of the star at an optical depth of $\tau_r = 2/3$. The assumptions used to arrive at this result is known as the Eddington approximation. The temperature as a function of optical depth is then

$$T^4 = \frac{3}{4} T_{\text{eff}}^4 \left(\tau_r + \frac{2}{3} \right). \tag{6.55}$$

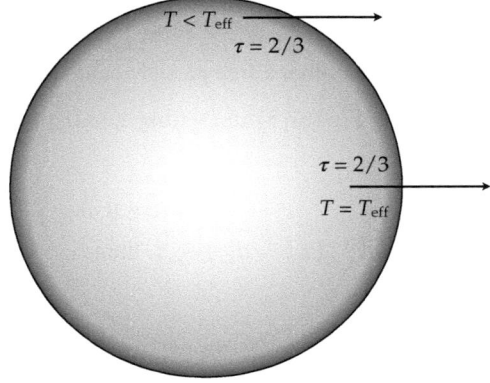

Fig. 6.7 The geometry showing the optical path of light leaving the surface of a star near the visible edge of the star disk. Note that the optical path is 2/3, but the radial optical depth, τ_r, is less than 2/3

Problem 6.3: Using Eq. (6.18) and setting $\tau_r = 0$, the specific intensity at the surface is given by

$$I_\lambda = -\frac{1}{\cos\theta}\int_0^\infty S_\lambda(t,\theta)e^{-t/\cos\theta}dt.$$

Show that if $S_\lambda(\tau_r) = a_\lambda + b_\lambda \tau$, then $I_\lambda = a_\lambda + b_\lambda \cos\theta$.

The major consequence here is that the light one sees from a star comes from an optical depth of 2/3. When one is looking at the edge of a star, the light comes from a smaller vertical depth and therefore from a region of the atmosphere that is at a lower temperature. This implies that the surface brightness of the star will be lower at the edges. This phenomenon is known as *limb darkening*. The appropriate geometry is shown in Fig. 6.7. The other consequence of Eq. (6.55) is that it was arrived at using the grey atmosphere. At frequencies where there is an atomic absorption line, then the physical depth is substantially smaller for an optical depth of 2/3, and so the light at absorption lines comes from a smaller τ_r than the light in the continuum. Therefore, it is at a lower temperature and is less luminous. This is why absorption lines appear to be darker.

Problems

6.1. Show that you can recover the expression for P_{rad} in Eq. (5.27) if you use B_λ, the blackbody expression for the specific intensity.

6.2. The relativistic momentum of a particle is $\gamma m v$ and the relativistic energy is $\gamma m c^2$, where $\gamma = \left(1 - v^2/c^2\right)^{-1/2}$. Use relativistic conservation of momentum and energy to show that a free particle cannot absorb a photon.

6.3. Using Eq. (6.18) and setting $\tau_r = 0$, the specific intensity at the surface is given by

$$I_\lambda = -\frac{1}{\cos\theta} \int_0^\infty S_\lambda(t,\theta) e^{-t/\cos\theta} dt.$$

Show that if $S_\lambda(\tau_r) = a_\lambda + b_\lambda \tau$, then $I_\lambda = a_\lambda + b_\lambda \cos\theta$.

Chapter 7
Nuclear Processes

The energy source that powers the luminosity of stars is nuclear fusion. Atomic nuclei consist of protons and neutrons that are bound tightly together in a volume a few femtometers across. The strong nuclear force that interacts with nucleons and binds them must be stronger than the repulsive electrostatic force that the protons feel in such close proximity to each other. The nuclear force cannot be long range or it would draw all nucleons in the universe together. A simplified cartoon of the potential felt by a proton in and near a nucleus is shown in Fig. 7.1. As a proton approaches it feels a repulsive electrostatic potential due to the net positive charge of the protons already in the nucleus. Once it gets close enough, the strong nuclear force takes over and the potential becomes negative and the proton becomes bound. If the new nucleus is less massive than iron, the rest mass energy of the new nucleus is lower than the combined rest masses of the incoming proton and the old nucleus. This is a direct result of the release of the binding energy.

7.1 Nuclear Fusion

In order for fusion to occur, two nuclei must approach one another close enough for the strong nuclear force to dominate over the electrostatic repulsion. Classically the average kinetic energy of a nucleus in a gas is given by $\frac{3}{2}kT$. In the absence of quantum effects, this energy would have to be equal to the electrostatic energy at separations on the order of a fermi (10^{-15} m); thus for two nuclei with atomic numbers $\mathscr{Z}_1$ and $\mathscr{Z}_2$ classical interactions would require temperatures given by

$$\frac{3}{2}kT = \frac{1}{4\pi\varepsilon_0}\frac{\mathscr{Z}_1\mathscr{Z}_2 e^2}{r} \tag{7.1}$$

in order for these two nuclei to fuse. Assuming that each nucleus is hydrogen, we find $T \sim 10^{10}$ K. This is about three orders of magnitude higher than typical core temperatures of solar-type stars, and the temperature only increases for heavier elements. Consequently, quantum tunneling must be important.

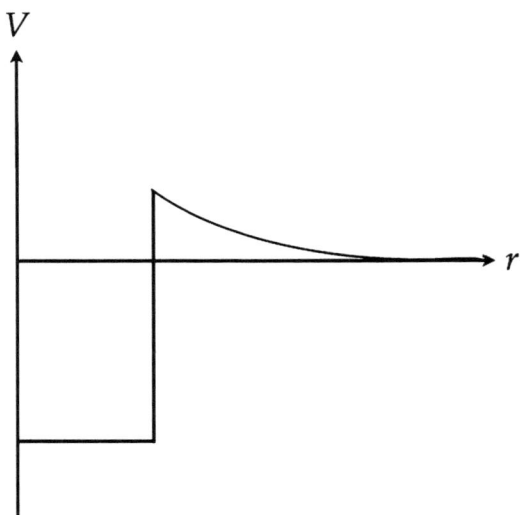

Fig. 7.1 Cartoon of the basic features of the nuclear potential. At large distances, the potential looks like a r^{-1} electrostatic repulsion. Within a few femtometers, the potential becomes attractive and goes negative. Bound nucleons will occupy discrete energy levels in the potential well

Remembering from our discussion of reaction rates in Chap. 4, we have the following expression for the number of reactions per unit volume per unit time for a given reaction:

$$r_{ix} = \int_0^\infty n_i n_x \sigma(E) v(E) \frac{n_E}{n} dE. \tag{7.2}$$

In this chapter, we will estimate the functional form of $\sigma(E)$. Since $\sigma(E)$ acts something like a physical area and since tunneling effects become important when the wavefunctions of the target and incident nuclei overlap, we can approximate the cross section by

$$\sigma(E) \propto \pi \lambda^2 = \pi \left(\frac{h}{p}\right)^2 \propto \frac{1}{E}, \tag{7.3}$$

where we have assumed that the typical size of the wavefunction is λ and that $E = p^2/2m$. The probability for tunneling to occur scales as $e^{-2\pi^2 U_c/E}$ where U_c is the Coulomb barrier height. The ratio of Coulomb barrier height to particle energy can be written as

$$\frac{U_c}{E} = \frac{\mathscr{Z}_1 \mathscr{Z}_2 e^2}{2\pi \varepsilon_0 h v}, \tag{7.4}$$

where we have assumed $r \sim \lambda \sim h/mv$, and so we have $\sigma(E) \propto e^{-b/\sqrt{E}}$, where b is a constant:

$$b = \frac{\pi \sqrt{m} \mathscr{Z}_1 \mathscr{Z}_2 e^2}{\sqrt{2}\varepsilon_0 h}. \tag{7.5}$$

7.1 Nuclear Fusion

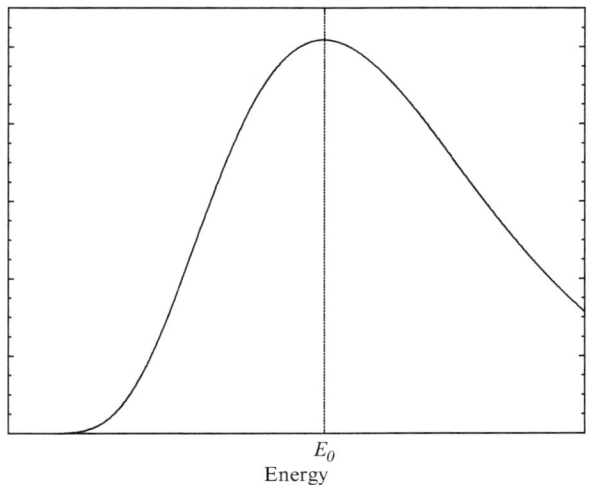

Fig. 7.2 The shape of the integrand in Eq. (7.8), assuming that $S(E)$ is a constant. The peak occurs at E_0 and indicates that a narrow range of energies contributes to the reaction rates for a given temperature T

We can combine these two arguments to obtain

$$\sigma(E) \propto \frac{e^{-b/\sqrt{E}}}{E}, \tag{7.6}$$

and we assume the proportionality "constant" is some slowly varying function of E, so

$$\sigma(E) = \frac{S(E)}{E} e^{-b/\sqrt{E}}. \tag{7.7}$$

Thus, the reaction rate can be expressed as

$$r_{ix} = \left(\frac{2}{kT}\right)^{3/2} \frac{n_i n_x}{\sqrt{m\pi}} \int_0^\infty S(E) e^{-b/\sqrt{E}} e^{-E/kT} dE, \tag{7.8}$$

where the first exponent in the integrand describes the tunneling probability and the second describes the high-energy tail of the Maxwell-Boltzmann distribution. Assuming $S(E)$ to be very nearly constant, the integrand is peaked at

$$E_0 = \left(\frac{bkT}{2}\right)^{2/3} \tag{7.9}$$

as can be seen in Fig. 7.2.

A nucleus has energy levels much like an atom does. At the energies corresponding to a bound energy state, $S(E)$ can increase dramatically. The increase in cross section comes from a resonance between the incident particle energy and the target nuclear energy levels. Consequently, these are called *resonance peaks*. This is similar to the opacity in a gas spiking at photon wavelengths that correspond to atomic transition energies; except in this case, it is the reaction rate that spikes.

At the temperatures and densities where reactions are occurring, reaction rates can be described as a power law centered on a particular temperature as

$$r_{ix} \simeq r_0 X_i X_x \rho^{\alpha'} T^\beta, \tag{7.10}$$

where r_0 is a constant specific to a given reaction, X_i and X_x are the mass fractions of the two particle species, and α' and β are determined from the expansion of the true reaction rate equation. For a two-body reaction, $\alpha' = 2$.

The amount of energy released per reaction is related to the difference in mass between the initial and final particles. For reactions of the type

$$I(\mathscr{A}_i, \mathscr{Z}_i) + J(\mathscr{A}_j, \mathscr{Z}_j) \rightleftharpoons K(\mathscr{A}_k, \mathscr{Z}_k) + L(\mathscr{A}_l, \mathscr{Z}_l), \tag{7.11}$$

the energy released is

$$Q_{ijk} = (\mathscr{M}_i + \mathscr{M}_j - \mathscr{M}_k - \mathscr{M}_l)c^2, \tag{7.12}$$

where $\mathscr{M}$ is the true mass of the nucleus. This value is usually different than the mass ($\mathscr{A} m_H$) that we have been using. The difference is called the *mass excess* and is usually written as an energy:

$$\Delta\mathscr{M}(I) = (\mathscr{M}_i - \mathscr{A}_i m_H)c^2. \tag{7.13}$$

Defining $\varepsilon_0 = Q_{ijk}$ for a given reaction, then

$$\varepsilon_{ix} = \left(\frac{\varepsilon_0}{\rho}\right) r_{ix} = \varepsilon_0' X_i X_x \rho^\alpha T^\beta, \tag{7.14}$$

where $\alpha = \alpha' - 1$ and ε_{ix} has units of W/kg. If the specific reactions involve neutrinos which can carry away energy without interacting with the star, then $\varepsilon_0 = Q_{ijk} - \varepsilon_0^\nu$. Thus, the value of q is obtained by summing over all reactions. Since most reactions are between two particles, this can be expressed as

$$q = q_0 \rho T^\beta \tag{7.15}$$

except in rare cases where more than two particles must collide for a reaction to proceed.

7.2 Hydrogen Burning

Although each step in the fusion burning process consists of the collision of two or three nuclei, many of the products of these collisions are unstable. Thus nuclear fusion proceeds along chains of many reactions until a stable end product is reached. In some chains, some nuclei serve as catalysts that shepherd the buildup of heavier nuclei. In this section we outline the basic reaction chains that operate at stages during the lifetimes of most stars. The notation we use in describing the components of each reaction is $^{\mathscr{A}}X$, where X is the chemical symbol for the element. Thus, X tells us the value of $\mathscr{Z}$.

7.2.1 The p-p Chains

The first reaction chain to occur in stars is the burning of hydrogen into helium. This is a process involving a chain of different reactions with the net result that $4H \rightarrow {}^4He$. The first link in this chain involves the collision of two protons:

$$p + p \rightarrow {}^2D + e^+ + \nu, \tag{7.16}$$

where one of the protons is converted into a neutron with the release of an antielectron (e^+) and a neutrino (ν). The product is a proton bound with a neutron in a nucleus known as a *deuteron*, given by the symbol ^2D. The deuteron interacts with another proton to form ^3He via

$$^2D + p \rightarrow {}^3He + \gamma, \tag{7.17}$$

where γ indicates an energetic photon. At this point in the chain, the ^3He nucleus can either interact with another ^3He nucleus that has been created by the same chain, or with a ^4He nucleus that is either primordial or has been produced in the stellar core. If it interacts with another ^3He nucleus, then

$$^3He + {}^3He \rightarrow {}^4He + 2p \tag{7.18}$$

completing what is known as the p-p I chain. The entire p-p I chain is shown in Fig. 7.3. The net result is

$$^1H + {}^1H + {}^1H + {}^1H \rightarrow {}^4He + 2e^+ + 2\nu + 2\gamma. \tag{7.19}$$

The positrons will annihilate electrons in the plasma and produce more photons which will scatter off of gas particles and heat the gas. The neutrinos will simply carry energy away from the star and not contribute to heating.

Fig. 7.3 The p-p I chain

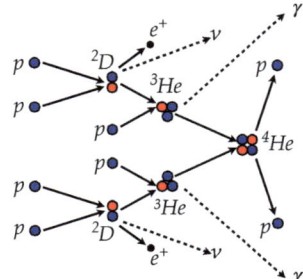

Fig. 7.4 The p-p II chain

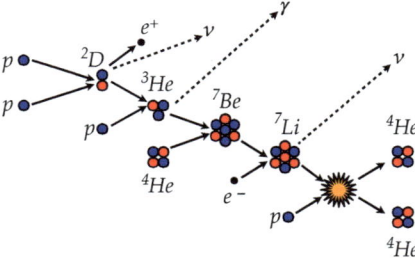

The other possible interaction of the ^{3}He involving a ^{4}He nucleus is the start of two additional proton–proton chains. The reaction results in

$$^3\text{He} + {}^4\text{He} \rightarrow {}^7\text{Be} + \gamma. \tag{7.20}$$

The ^{7}Be nucleus can then interact with an electron in the plasma to branch into the p-p II chain:

$$^7\text{Be} + e^- \rightarrow {}^7\text{Li} + \nu \tag{7.21}$$

followed by

$$^7\text{Li} + p \rightarrow 2\,{}^4\text{He}. \tag{7.22}$$

The p-p II chain is shown in Fig. 7.4. If the ^{7}Be nucleus interacts with a proton in the plasma, it follows the p-p III chain:

$$^7\text{Be} + p \rightarrow {}^8\text{B} + \gamma \tag{7.23}$$

followed by

$$^8\text{B} \rightarrow {}^8\text{Be} + e^+ + \nu, \tag{7.24}$$

$$^8\text{Be} \rightarrow 2\,{}^4\text{He}. \tag{7.25}$$

Figure 7.5 shows the p-p III chain.

7.2 Hydrogen Burning

Fig. 7.5 The p-p III chain

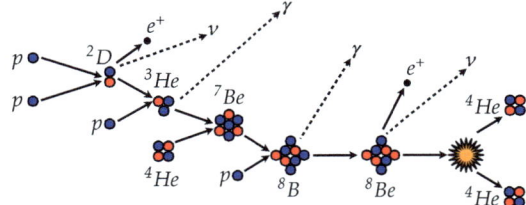

The probability of each of these chains depends upon the temperature, density, and composition of the plasma. In typical conditions in the solar core, the p-p I chain occurs 69% of the time and the p-p II and III chains occur the remaining 31%. Of the last 31%, the p-p II chain occurs 99.7% of the time and the p-p III 0.3%.

The energy released by the reaction is 26.73 MeV, but some of this energy is carried away by neutrinos which do not contribute to heating the star. The $p + p \to {}^2D + e^+ + \nu$ reaction loses 0.26 MeV to neutrinos, while the ${}^8B \to {}^8Be + e^+ + \nu$ reaction loses up to 7.2 MeV. Because the branching ratio for the p-p III chain is so small, the average energy released per ${}^4He$ created is about 26 MeV.

The slowest reaction in the chain governs the rate of energy release, and for temperatures around 1.5×10^7 K, the power law form can be written

$$q_{pp} = q_{0,pp} \rho X^2 T_6^4, \tag{7.26}$$

where $T_6 = T/10^6$ K and $q_{0,pp} = 1.08 \times 10^{-12}$ W m^3/kg^2.

Problem 7.1: The source of the luminosity of the sun is the fusion of 4 ${}^1H$ nuclei to form one ${}^4He$ nucleus. The energy from each reaction comes from the conversion of mass to energy via $E = mc^2$. Using the luminosity of the sun, calculate the mass loss rate (in kg/s) due to nuclear fusion.

Problem 7.2: In the p-p chain, the net energy released for each ${}^4He$ created is 26 MeV. Assuming that the entire luminosity of the sun is generated by the p-p chain, calculate the mass of ${}^1H$ that is converted to ${}^4He$ (in kg/s).

7.2.2 The CNO Cycle

The initial composition of most stars will also contain a mass fraction of C, N, and O nuclei. At higher temperatures, these nuclei can act as catalysts for the burning of

Fig. 7.6 The CNO I cycle. The cycle starts at the top with a ^{12}C and four protons are added, finishing with a ^{4}He nucleus and returning to the ^{12}C nucleus

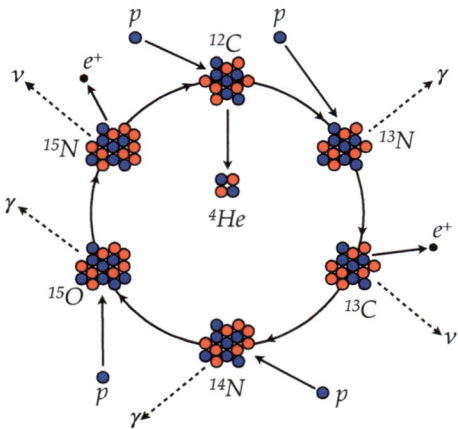

hydrogen into helium. The CNO cycle also comes in two chains. The dominant one follows:

$$^{12}C + {}^1H \rightarrow {}^{13}N + \gamma, \tag{7.27}$$
$$^{13}N \rightarrow {}^{13}C + e^+ + \nu, \tag{7.28}$$
$$^{13}C + {}^1H \rightarrow {}^{14}N + \gamma, \tag{7.29}$$
$$^{14}N + {}^1H \rightarrow {}^{15}O + \gamma, \tag{7.30}$$
$$^{15}O \rightarrow {}^{15}N + e^+ + \nu, \tag{7.31}$$
$$^{15}N + {}^1H \rightarrow {}^{12}C + {}^4He, \tag{7.32}$$

Note that the net result is $4\,{}^1H \rightarrow {}^4He$ and we get our ^{12}C back. This is the CNO I cycle, and it is shown in Fig. 7.6.

Sometimes the last reaction fails to generate ^{4}He and instead follows:

$$^{15}N + {}^1H \rightarrow {}^{16}O + \gamma, \tag{7.33}$$
$$^{16}O + {}^1H \rightarrow {}^{17}F + \gamma, \tag{7.34}$$
$$^{17}F \rightarrow {}^{17}O + e^+ + \nu, \tag{7.35}$$
$$^{17}O + {}^1H \rightarrow {}^{14}N + {}^4He. \tag{7.36}$$

In this case, the net result is that $6\,{}^1H + {}^{12}C \rightarrow {}^4He + {}^{14}N$, but we can also consider the reaction chain to start at $^{14}N + {}^1H \rightarrow {}^{15}O + \gamma$ and the result is $4\,{}^1H \rightarrow {}^4He$ and we get our ^{14}N back, as shown in Fig. 7.7. Note that the only difference between the CNO II cycle and the CNO I cycle is the addition of a proton and a neutron to each nucleus.

7.3 Burning Heavier Nuclei

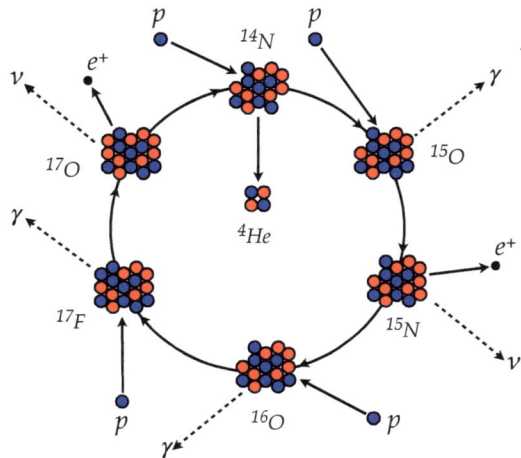

Fig. 7.7 The CNO II cycle. The cycle starts at the top with a ^{4}N and four protons are added, finishing with a ^{4}He nucleus and returning to the ^{14}N nucleus

Because there are more neutrinos generated in the CNO cycle, the net energy produced per helium nucleus is a little lower at ~ 25 MeV. Again, at temperatures around 1.5×10^7 K, this cycle generates

$$q_{CNO} = q_{0,CNO} X X_{CNO} \rho T_6^{19.9} \tag{7.37}$$

with $q_{0,CNO} = 8.24 \times 10^{-31}$ W m^3 kg^{-2} and the combined mass fraction of C, N, and O is given by X_{CNO}. Note that although $q_{0,CNO}$ is much lower than $q_{0,pp}$ and X_{CNO} is much lower than X, the steep T-dependence means that this cycle can be the dominant energy generation mechanism in the more massive stars that have higher central temperatures.

Note that in all of these hydrogen burning cycles, X will decrease as the hydrogen is converted into helium. Eventually, these cycles will not be able to supply the energy needed to support the star. At this point, the core will begin to contract and if the temperature and density become high enough, helium will eventually begin to fuse .

7.3 Burning Heavier Nuclei

When helium fuses, the first reaction

$$^4\text{He} + ^4\text{He} \rightleftharpoons ^8\text{Be} \tag{7.38}$$

is very unstable and short-lived, with the ^{8}Be nucleus decaying back into two ^{4}He nuclei in about 10^{-16} s. Consequently, the reaction can only proceed if another alpha particle strikes the ^{8}Be nucleus almost simultaneously as it is formed. This reaction results in

$$^4\text{He} + {}^8\text{Be} \to {}^{12}\text{C} + \gamma. \tag{7.39}$$

The cross section for this reaction is quite large due to the fact that the combined energies of the He and Be nuclei are close to an energy level of 12C, and so there is a resonance. Nonetheless, the effective reaction is a three-body process and so the energy production rate depends on the mass fraction of helium *cubed* and the density *squared*:

$$q \simeq q_{0,3\alpha} \rho^2 Y^3 T_8^{41.0}, \tag{7.40}$$

where $T_8 = T/10^8$ K and $q_{0,3\alpha} = 3.96 \times 10^{-18}$ W m^6 kg^{-3}. Once a sufficient number of ^{12}C nuclei have been built up, these can interact with ^{4}He nuclei to generate ^{16}O via

$$^{12}\text{C} + {}^4\text{He} \to {}^{16}\text{O} + \gamma. \tag{7.41}$$

These oxygen nuclei can combine with helium to produce neon:

$$^{16}\text{O} + {}^4\text{He} \to {}^{20}\text{Ne} + \gamma. \tag{7.42}$$

At the temperatures and densities maintained during helium burning the Coulomb barrier is too high for neon to fuse. Once the helium fuel is exhausted, the core will again collapse and heat up. If the star is sufficiently massive, then carbon burning can be ignited. Further fusion reactions are not very well modeled by temperature power laws, and so we only list the reactions and their threshold temperatures.

Carbon burning begins at about 6×10^8 K. Some of these reactions absorb energy, and are marked with an asterisk:

$$^{12}\text{C} + {}^{12}\text{C} \to {}^{24}\text{Mg} + \gamma \tag{7.43}$$

$$* \to {}^{23}\text{Mg} + n \tag{7.44}$$

$$\to {}^{23}\text{Na} + {}^1\text{H} \tag{7.45}$$

$$\to {}^{20}\text{Ne} + {}^4\text{He} \tag{7.46}$$

$$* \to {}^{16}\text{O} + 2\,{}^4\text{He}. \tag{7.47}$$

Oxygen burning does not begin until $T \sim 10^9$ K. Again some reactions are endothermic:

$$^{16}\text{O} + {}^{16}\text{O} \to {}^{32}\text{S} + \gamma \tag{7.48}$$

$$* \to {}^{31}\text{S} + n \tag{7.49}$$

$$\to {}^{31}\text{P} + {}^1\text{H} \tag{7.50}$$

$$\to {}^{28}\text{Si} + {}^4\text{He} \tag{7.51}$$

$$* \to {}^{24}\text{Mg} + 2\,{}^4\text{He}. \tag{7.52}$$

By the time temperatures have reached the oxygen burning threshold, nearly all ^{12}C nuclei have been burned, so ^{12}C + ^{16}O reactions are negligible.

Problem 7.3: In a fully convective core, the material within the core is continually mixed so that the mass fractions are uniform throughout the core. A star has a fully convective core that is pure ^{1}H and has a mass of 0.1 M$_\odot$.

(a) Assume that all of the hydrogen is burned to produce a core of pure ^{4}He. What is the mass of the resulting helium core? What is the total amount of energy released during the hydrogen burning phase?

(b) Assume that all of the helium in the helium core is subsequently burned to produce a core of pure ^{12}C. What is the mass of the resulting carbon core? What is the total amount of energy released during the helium burning phase?

By the time we reach temperatures where ^{28}Si + ^{28}Si reactions could occur, the ambient photons are energetic enough to begin reversing the fusion process through a process known as *photodisintegration*. In this case heavy nuclei are broken up into lighter nuclei which can then overcome the Coulomb barrier more easily, and so nuclei heavier than ^{28}Si are built up by reactions with lighter nuclei that arise from photodisintegration. A typical chain might flow like

$$^{28}\text{Si} + {}^{4}\text{He} \rightleftharpoons {}^{32}\text{S} + \gamma, \tag{7.53}$$

$$^{32}\text{S} + {}^{4}\text{He} \rightleftharpoons {}^{36}\text{Ar} + \gamma, \tag{7.54}$$

$$\cdots$$

$$^{32}\text{Cr} + {}^{4}\text{He} \rightleftharpoons {}^{56}\text{Ni} + \gamma. \tag{7.55}$$

Any reactions that produce elements higher than ^{56}Fe are endothermic. The reactions described above are listed in Table 7.1 along with threshold temperatures and typical duration for a 20 M$_\odot$ star. The nuclear binding energy per nucleon is shown in Fig. 7.8. The rest masses of many elements can be found in Appendix B.

7.4 Neutron Capture Processes

Another process that can build up heavy nuclei arises from free neutrons that are generated by processes such as C, O, or Si burning. These neutrons are not hindered by the Coulomb barrier, so the reaction rates are determined by the number of neutrons. The reactions are of two types:

Neutron capture: $I(\mathscr{A}, \mathscr{Z}) + n \rightarrow I(\mathscr{A}+1, \mathscr{Z})$
Beta decay: $I(\mathscr{A}, \mathscr{Z}) \rightarrow J(\mathscr{A}, \mathscr{Z}+1) + e^{-} + \bar{\nu}.$

Table 7.1 Common nuclear processes in stars

Nuclear fuel	Process	T_{thresh} 10^6 K	Products	Energy per nucleon (MeV)	Duration in a 20 $M_\odot$ star
H	$p-p$	~4	He	6.55	10^7 year
H	CNO	15	He	6.25	10^7 year
He	3α	100	C, O	0.61	10^6 year
C	C+C	600	O, Ne, Na, Mg	0.54	300 year
O	O+O	1,000	Mg, S, P, Si	~0.3	200 d
Si	Nuc. eq.	3,000	Co, Fe, Ni	<0.18	2 d

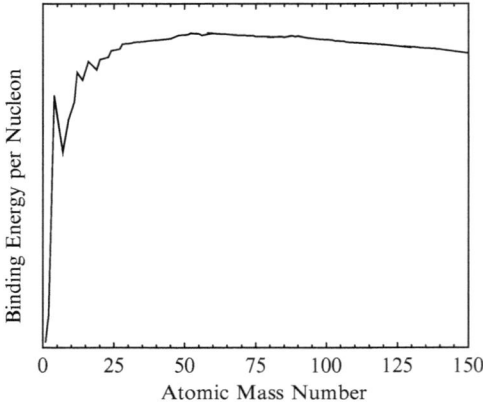

Fig. 7.8 The binding energy per nucleon as a function of atomic mass number (in arbitrary units). As you move from lower to higher binding energy per nucleon, energy is released in the reaction, but if you move from higher to lower, then the reaction is endothermic

The timescales of β-decay are set by the weak force and are independent of local temperatures and densities. Neutron capture rates are set by the local density and may vary by temperature as neutrons may participate in resonant reactions. If neutron capture rates are slower than the β-decay rates, the process is called an *s-process* (for "slow"). If the neutron capture rates are faster than β-decay rates, then the process is called an *r-process* (for "rapid"). Some nuclei are built up through both r- and s-processes, while others are only produced by one process. These reactions also tend to be endothermic.

At very high energies and densities, a number of inverse reactions occur that eventually return the core to a collection of protons, neutrons, and electrons. These will be discussed in more detail as we study neutron stars and supernovae.

Problems

7.1. The source of the luminosity of the sun is the fusion of 4 ^{1}H nuclei to form one ^{4}He nucleus. The energy from each reaction comes from the conversion of mass to energy via $E = mc^2$. Using the luminosity of the sun, calculate the mass loss rate (in kg/s) due to nuclear fusion.

7.2. In the p-p chain, the net energy released for each ^{4}He created is 26 MeV. Assuming that the entire luminosity of the sun is generated by the p-p chain, calculate the mass of ^{1}H that is converted to ^{4}He (in kg/s).

7.3. In a fully convective core, the material within the core is continually mixed, so that the mass fractions are uniform throughout the core. A star has a fully convective core that is pure ^{1}H and has a mass of 0.1 $M_\odot$.

(a) Assume that all of the hydrogen is burned to produce a core of pure ^{4}He. What is the mass of the resulting helium core? What is the total amount of energy released during the hydrogen burning phase?
(b) Assume that all of the helium in the helium core is subsequently burned to produce a core of pure ^{12}C. What is the mass of the resulting carbon core? What is the total amount of energy released during the helium burning phase?

Part III
Stellar Models

With the equations of stellar evolution in hand and an understanding of the physical processes at play, we can begin modeling the life cycles of stars. Although numerical methods are necessary for accurate descriptions, many general properties and stages of stellar life can be understood through analytical means. In this section, we develop some analytical models and descriptions of the many stages of stellar life.

Chapter 8
Simple Stellar Models

Now that we have an understanding of the physics behind both the equations of stellar structure and their auxiliary functions, we are in a position to describe some simple static models of stellar structure. Setting all time derivatives to zero gives the static structure equations:

$$\frac{dP}{dm} = -\frac{Gm}{4\pi r^4}, \quad (8.1)$$

$$\frac{dr}{dm} = \frac{1}{4\pi r^2 \rho}, \quad (8.2)$$

$$\frac{dT}{dm} = -\frac{3}{4ac}\frac{\kappa}{T^3}\frac{F}{(4\pi r^2)^2}, \quad (8.3)$$

$$\frac{dF}{dm} = q. \quad (8.4)$$

The auxiliary equations are

$$P = \frac{\mathcal{R}}{\mu_I}\rho T + P_e + \frac{1}{3}aT^4, \quad (8.5)$$

$$\kappa = \kappa_0 \rho^a T^b, \quad (8.6)$$

$$q = q_0 \rho^\alpha T^\beta. \quad (8.7)$$

8.1 Polytropes

We begin by looking at solutions that can be obtained analytically for a set of somewhat reasonable assumptions. First, we note that if there existed a simple relationship between P and ρ that was independent of T, then the first two equations could be solved without reference to the second two equations. We have seen some of these equations of state already as

$$P = K\rho^\gamma, \tag{8.8}$$

which describe degenerate gases. These are known as polytropic equations of state and γ is related to the *polytropic index n* by $\gamma = (n+1)/n$. Stars whose structure is determined by a polytropic equation of state are called *polytropes*. Polytropes can be used to model degenerate stars such as neutron stars or white dwarfs.

We start solving for the polytropic model by manipulating the dynamical equation as follows:

$$\frac{r^2}{\rho}\frac{dP}{dr} = -Gm, \tag{8.9}$$

$$\frac{d}{dr}\left(\frac{r^2}{\rho}\frac{dP}{dr}\right) = -G\frac{dm}{dr}, \tag{8.10}$$

$$\frac{1}{r^2}\frac{d}{dr}\left(r^2\left(\frac{1}{\rho}\frac{dP}{dr}\right)\right) = -4\pi G\rho. \tag{8.11}$$

Note that the left-hand side is suspiciously like the radial part of the Poisson equation. This is not surprising, when we note that Eq. (8.1) can also be written as

$$\frac{dP}{dr} = \frac{dm}{dr}\frac{dP}{dm} = -\frac{Gm}{r^2}\rho = \frac{d\phi}{dr}\rho, \tag{8.12}$$

where ϕ is the gravitational potential at the surface of m. Therefore,

$$\frac{1}{\rho}\frac{dP}{dr} = \frac{d\phi}{dr}, \tag{8.13}$$

and Eq. (8.11) becomes

$$\frac{1}{r^2}\frac{d}{dr}\left(r^2\frac{d\phi}{dr}\right) = \nabla^2\phi = -4\pi G\rho. \tag{8.14}$$

Using the polytrope equation of state, we can recast Eq. (8.11) into a differential equation governing ρ:

$$\frac{1}{r^2}\frac{d}{dr}\left(r^2\left(\frac{1}{\rho}K\gamma\rho^{\gamma-1}\frac{d\rho}{dr}\right)\right) = -4\pi G\rho, \tag{8.15}$$

or

$$\frac{\gamma K}{4\pi G}\frac{1}{r^2}\frac{d}{dr}\left(r^2\rho^{\gamma-2}\frac{d\rho}{dr}\right) = -\rho. \tag{8.16}$$

8.1 Polytropes

We now set about rewriting this equation in terms of dimensionless quantities. This is a standard approach to obtaining solutions to many differential equations. Let $\rho = \rho_c \theta^n$ where θ is a dimensionless function of r, such that $0 < \theta(r) < 1$ and $\rho_c =$ central density. Substituting gives

$$\frac{\gamma K}{4\pi G} \frac{1}{r^2} \frac{d}{dr}\left(r^2 \rho_c^{\gamma-2} \theta^{(\gamma-2)n} \rho_c n \theta^{n-1} \frac{d\theta}{dr}\right) = \rho_c \theta^n, \tag{8.17}$$

but

$$(\gamma-2)n = \left(\frac{n+1}{n} - 2\right)n = 1 - n, \tag{8.18}$$

so

$$\frac{n\gamma K}{4\pi G} \rho_c^{\gamma-1} \frac{1}{r^2} \frac{d}{dr}\left(r^2 \frac{d\theta}{dr}\right) = -\rho_c \theta^n, \tag{8.19}$$

or

$$\left[\frac{(n+1)K}{4\pi G} \rho_c^{\gamma-2}\right] \frac{1}{r^2} \frac{d}{dr}\left(r^2 \frac{d\theta}{dr}\right) = -\theta^n. \tag{8.20}$$

Since the term in brackets is a constant with units of area, we write it as

$$\left[\frac{(n+1)K}{4\pi G} \rho_c^{\gamma-2}\right] = \alpha^2, \tag{8.21}$$

where α is a length. We can now introduce a dimensionless variable ξ such that $r = \alpha\xi$, and then the equation reads

$$\frac{1}{\xi^2} \frac{d}{d\xi}\left(\xi^2 \frac{d\theta}{d\xi}\right) = -\theta^n. \tag{8.22}$$

This is known as the *Lane–Emden equation*.

We need two boundary conditions in order to solve the Lane–Emden equation. For the first boundary condition, we note that $\theta(0) = 1$ in order for $\rho(0) = \rho_c$. For the second boundary condition, we expand the derivatives in Eq. (8.22) to find

$$\frac{1}{\xi^2} \frac{d}{d\xi}\left(\xi^2 \frac{d\theta}{d\xi}\right) = \frac{2}{\xi} \frac{d\theta}{d\xi} + \frac{d^2\theta}{d\xi^2} = -\theta^n. \tag{8.23}$$

At $\xi = 0$, the right-hand side becomes $-\theta^n = -1$, but the left-hand side has a division by 0 in the first term. Either we force $d^2\theta/d\xi^2$ to be equally undefined but with opposite sign (which is unphysical), or we require $d\theta/d\xi = 0$ at $\xi = 0$. This is our second boundary condition. Therefore, our two boundary conditions are

$$\theta(0) = 1 \quad \text{and} \quad \left.\frac{d\theta}{d\xi}\right|_0 = 0. \tag{8.24}$$

Since the density cannot be negative, we define the first zero of $\theta(\xi)$ to correspond to the radius of the star, so $\theta(\xi_1) = 0 \Rightarrow R = \alpha\xi_1$.

In order to relate the solution of the Lane–Emden equation to the total mass, we note that

$$M = \int_0^R 4\pi r^2 \rho \, dr = 4\pi\alpha^3 \rho_c \int_0^{\xi_1} \xi^2 \theta^n \, d\xi$$

$$= -4\pi\alpha^3 \rho_c \int_0^{\xi_1} \frac{d}{d\xi}\left(\xi^2 \frac{d\theta}{d\xi}\right) d\xi$$

$$= -4\pi\alpha^3 \rho_c \xi_1^2 \left(\frac{d\theta}{d\xi}\right)\bigg|_{\xi_1}. \tag{8.25}$$

Noting that $R = \alpha\xi_1 \Rightarrow \alpha = R/\xi_1$, so

$$M = -4\pi\left(\frac{R}{\xi_1}\right)^3 \rho_c \xi_1^2 \left(\frac{d\theta}{d\xi}\right)\bigg|_{\xi_1}$$

$$= \frac{4}{3}\pi R^3 \rho_c \left[\frac{-3}{\xi_1}\left(\frac{d\theta}{d\xi}\right)\bigg|_{\xi_1}\right] \tag{8.26}$$

or

$$\frac{M}{\left(\frac{4}{3}\pi R^3\right)} = \bar{\rho} = \rho_c \left[\frac{-3}{\xi_1}\left(\frac{d\theta}{d\xi}\right)\bigg|_{\xi_1}\right]. \tag{8.27}$$

The relation between average density, $\bar{\rho}$, and central density, ρ_c, is

$$\rho_c = D_n \bar{\rho}, \tag{8.28}$$

where

$$D_n = \left[\frac{-3}{\xi_1}\left(\frac{d\theta}{d\xi}\right)\bigg|_{\xi_1}\right]^{-1}. \tag{8.29}$$

The mass equation also provides a relation between mass and radius:

$$\left[\frac{GM}{M_n}\right]^{n-1}\left[\frac{R}{R_n}\right]^{3-n} = \frac{[(n+1)K]^n}{4\pi G}, \tag{8.30}$$

where $M_n = -\xi_1^2 \, (d\theta/d\xi)|_{\xi_1}$ and $R_n = \xi_1$. Note that there are two special cases: $n = 1$ and $n = 3$.

For $n = 1$ polytropes, we find that

$$\left[\frac{R}{R_2}\right]^2 = \frac{K}{2\pi G}, \tag{8.31}$$

8.1 Polytropes

and so only one value of R will satisfy hydrostatic equilibrium for a given value of K. Note that the equation of state for $n = 1$ is

$$P = K\rho^2. \tag{8.32}$$

For $n = 3$, then

$$\left[\frac{GM}{M_3}\right]^2 = \frac{16K^3}{\pi G}, \tag{8.33}$$

and there is only one possible mass that will satisfy hydrostatic equilibrium for a given K. For $n = 3$, the equation of state has $\gamma = 4/3$ and describes a fully relativistic degenerate gas. Thus, a fully relativistic degenerate gas defines a limiting mass for an equation of state given by K.

Between the limiting values of $n = 1$ and $n = 3$, we have

$$R^{3-n} \propto \frac{1}{M^{n-1}}. \tag{8.34}$$

In the case of a nonrelativistic degenerate gas,

$$P = K\rho^{5/3} \tag{8.35}$$

so $n = 1.5$. The Lane–Emden equation cannot be solved analytically for this value of n, but we can determine the mass-radius relation:

$$R \propto \frac{1}{M^{1/3}}, \tag{8.36}$$

so the radius decreases with increasing mass. Thus, the density obeys

$$\rho \propto \frac{M}{R^3} \propto M^2. \tag{8.37}$$

Therefore, the density increases as the square of the mass. Recalling that the Fermi energy increases with the density,

$$E_F = \frac{\hbar^2}{2m_e}\left[3\pi^2 \left(\frac{\mathscr{Z}}{\mathscr{A}}\right)\frac{\rho}{m_\text{H}}\right]^{2/3}, \tag{8.38}$$

we see that the electrons will eventually become relativistic and the equation of state will smoothly change over to

$$P = K\rho^{4/3} \tag{8.39}$$

with increasing mass. This is a polytrope with $n = 3$, which has only one value of the mass for hydrostatic equilibrium. This will be the maximum mass allowed for a star to be supported by degenerate electron pressure.

Stars supported by degenerate electron pressure are known as *white dwarfs*. When the maximum mass is exceeded, one possible outcome is a star supported by degenerate neutron pressure. These are known as *neutron stars*. Returning to the equation for a fully relativistic degenerate electron gas, we can find this maximum mass for different compositions. We have

$$P = \frac{hc}{8}\left(\frac{3}{\pi}\right)^{1/3}\frac{1}{m_H^{4/3}}\left(\frac{\rho}{\mu_e}\right)^{4/3}, \tag{8.40}$$

so

$$K = \frac{hc}{8}\left(\frac{3}{\pi}\right)^{1/3}\left(\frac{1}{m_H\mu_e}\right)^{4/3} \tag{8.41}$$

and

$$M = 4\pi M_3 \left(\frac{K}{4\pi G}\right)^{3/2}. \tag{8.42}$$

Numerical solutions of the $n = 3$ polytrope give $M_3 = 2.02$, and so

$$M = 4\pi(2.02)\left(\frac{1}{4\pi G}\right)^{3/2}\left(\frac{hc}{8}\right)^{3/2}\left(\frac{3}{\pi}\right)^{1/2}\frac{1}{m_H^2}\frac{1}{\mu_e^2} \tag{8.43}$$

$$\simeq 5.83\mu_e^{-2}M_\odot. \tag{8.44}$$

Recall from Chap. 5 that

$$\mu_e^{-1} = X + \frac{1}{2}Y + (1-X-Y)\left\langle\frac{\mathscr{Z}}{\mathscr{A}}\right\rangle. \tag{8.45}$$

For a hydrogen white dwarf, $\mu_e = 1$ and the maximum mass is $5.83\,M_\odot$. However, this is an unphysical situation since the pressures and densities attained in any reasonable evolution will ignite the p–p chain before a relativistic degenerate state is reached. For He, C, O, and Ne, $\langle\mathscr{Z}/\mathscr{A}\rangle = \frac{1}{2}$, and so $\mu_e = 2$, and the maximum mass is $1.46\,M_\odot$. These are reasonable masses, and all known white dwarfs have masses below this value. Some binary evolution combined with mass transfer to the white dwarf is necessary to achieve the maximum mass. For single star evolution, the final end product is an iron core of $^{56}_{26}$Fe, and so $\langle\mathscr{Z}/\mathscr{A}\rangle = 0.464$, so the maximum mass is $1.26\,M_\odot$. This explains that fact that there are neutron stars with masses around $1.3\,M_\odot$, even though the maximum mass of a white dwarf is greater than this value. The maximum mass for He, C, O, and Ne white dwarfs is known as the *Chandrasekhar Mass*.

8.2 Polytrope Solutions

The simplest analytic solution to the Lane–Emden equation is the case of $n = 0$. In this case,

$$\frac{1}{\xi^2}\frac{d}{d\xi}\left(\xi^2 \frac{d\theta}{d\xi}\right) = -1, \tag{8.46}$$

so

$$\frac{d}{d\xi}\left(\xi^2 \frac{d\theta}{d\xi}\right) = -\xi^2. \tag{8.47}$$

This can be directly integrated to obtain

$$\xi^2 \frac{d\theta}{d\xi} = -\frac{1}{3}\xi^3 + C_1. \tag{8.48}$$

Using the first boundary condition,

$$\left.\frac{d\theta}{d\xi}\right|_0 = 0, \tag{8.49}$$

we find $C_1 = 0$. Rearranging and integrating a second time give

$$\frac{d\theta}{d\xi} = -\frac{1}{3}\xi \implies \theta = -\frac{1}{6}\xi^2 + C_2. \tag{8.50}$$

The second boundary condition requires $\theta(0) = 1$, so $C_2 = 1$. Thus, the solution is:

$$\theta = 1 - \frac{1}{6}\xi^2, \tag{8.51}$$

and the zero is at $\xi_1 = \sqrt{6}$, so $R = \alpha \xi_1 \implies \alpha = R/\sqrt{6}$. The total mass is then given by

$$M = -4\pi \left(\frac{R}{\sqrt{6}}\right)^3 \rho_c \left(\sqrt{6}\right)^2 \left(-\frac{1}{3}\sqrt{6}\right)$$

$$= \frac{4}{3}\pi R^3 \rho_c. \tag{8.52}$$

Thus, the $n = 0$ case is a constant density solution. (Just because it is solvable, doesn't mean it is an interesting case.) Note that this should have been expected since $n = 0$ corresponds to $\gamma =$ undefined. Thus, $P = K\rho^\gamma$ is undefined and P must therefore be independent of ρ.

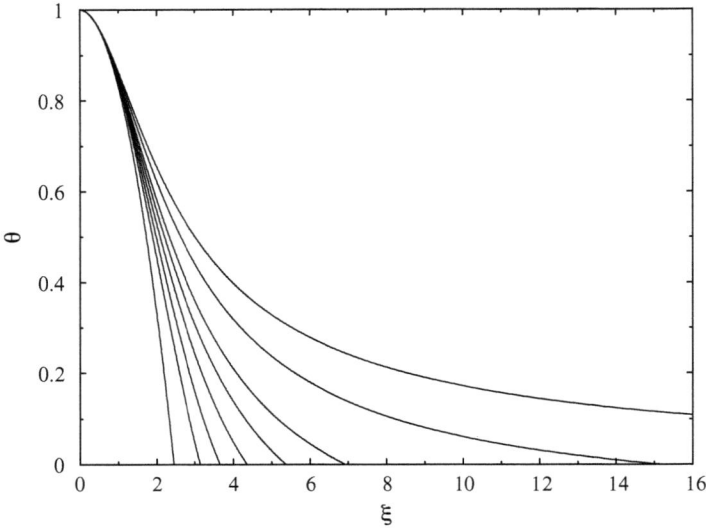

Fig. 8.1 Numerical solutions to the Lane–Emden equation for values of $n = 0, 1, 1.5, 2, 2.5, 3, 4,$ and 5. The solution for $n = 5$ asymptotically approaches 0

The Lane–Emden equation can be solved analytically for two other cases, $n = 1$ and $n = 5$. For all other cases, one must use numerical methods. One of the simplest approaches is to increment ξ by small steps $d\xi$, so

$$\theta_{i+1} = \theta_i + \left(\frac{d\theta}{d\xi}\right)_i d\xi, \tag{8.53}$$

where

$$\left(\frac{d\theta}{d\xi}\right)_{i+1} = \left(\frac{d\theta}{d\xi}\right)_i + \left(\frac{d^2\theta}{d\xi^2}\right)_i d\xi = \left(\frac{d\theta}{d\xi}\right)_i - \left(\frac{2}{\xi}\frac{d\theta}{d\xi} + \theta^n\right)_i d\xi. \tag{8.54}$$

Starting with the boundary condition, we start from the center of the star and work our way out toward the surface, stopping the computation when $\theta < 0$. The results of this computation are shown in Fig. 8.1. The important values for a range of polytropic indices are given in Table 8.1.

Problem 8.1: Solve the Lane–Emden equation for $n = 1$. Calculate the mass of the star in terms of the central density ρ_c.

8.3 The Eddington Standard Model

Table 8.1 Numerically computed values of R_n, M_n, and D_n for polytropes with values of n between 0 and 5

n	R_n	M_n	D_n
0	2.45	4.89	1.00
1	3.14	3.14	3.29
1.5	3.65	2.71	5.98
2	4.35	2.41	11.38
2.5	5.36	2.19	23.44
3	6.90	2.02	54.21
4	15.00	1.80	625.00
5	∞	–	–

8.3 The Eddington Standard Model

Using a polytropic equation of state, we can also obtain a simplified model of a nondegenerate star. In this model, known as the Eddington Standard Model, we assume only that the gas pressure is a constant fraction of the total pressure throughout the star. This implies that β is constant. We can write the total pressure in terms of the radiation pressure, so

$$P = \frac{P_{\text{gas}}}{\beta} = \frac{P_{\text{rad}}}{1-\beta}, \tag{8.55}$$

or

$$\frac{\mathcal{R}}{\beta\mu}\rho T = \frac{aT^4}{3(1-\beta)}. \tag{8.56}$$

Because β is constant, we can obtain a relation between T and ρ throughout the star:

$$T = \left[\frac{3\mathcal{R}(1-\beta)}{a\mu\beta}\right]^{1/3} \rho^{1/3}, \tag{8.57}$$

and the ideal gas law goes to a polytropic equation of state of index $n = 3$:

$$P = \frac{\mathcal{R}}{\mu}\rho T = \left[\frac{3\mathcal{R}^4(1-\beta)}{a\mu^4\beta^4}\right]^{1/3} \rho^{4/3}. \tag{8.58}$$

From Eq. (8.33), we can see that the mass of the star is uniquely determined by the constant in the equation of state. Therefore,

$$M = 4\pi M_3 \left(\frac{K}{\pi G}\right)^{3/2} = 4\pi M_3 \left(\frac{1}{\pi G}\right)^{3/2} \left[\frac{3\mathcal{R}^4(1-\beta)}{a\mu^4\beta^4}\right]^{1/2}. \tag{8.59}$$

This can be rewritten as an equation giving β as a function of μ and M:

$$(1-\beta) = \mu^4 M^2 \beta^4 \left[\frac{aG^3}{48\pi M_3^2 \mathcal{R}^4}\right] \simeq 0.03 \left[\frac{M}{M_\odot}\right]^2 \mu^4 \beta^4. \tag{8.60}$$

Problem 8.2: The solar value of the mean molecular weight is 0.61. Use this to determine β for the Eddington Standard Model of the sun.

Problem 8.3: Using the value of β found in Problem 8.2 and the solutions for the $n = 3$ polytrope, determine the central density and central temperature of the sun in the Eddington Standard Model.

Problem 8.4: Numerically solve the $n = 3$ polytrope equation using solar values to determine the radius at which the temperature is high enough for the p–p chain to be operating ($T_6 = 15$).

8.4 The Eddington Luminosity

If the radiation pressure dominates the gas pressure (as can occur in the atmospheres of stars or accretion disks), then

$$P \simeq P_{\text{rad}} \quad \text{and} \quad \frac{dP}{dr} \simeq \frac{dP_{\text{rad}}}{dr} = \frac{-\kappa \rho F}{4\pi c r^2}. \tag{8.61}$$

This pressure must be counterbalanced by the gravitational force if the star is to be in hydrostatic equilibrium. Consider starting at the surface of the star where $P = 0$. If the gradient of the radiation pressure is steeper than the gradient of the hydrostatic pressure, then the star is unstable and the outer layers will be pushed off. This is equivalent to stating that the star will be stable if

$$\left|\frac{dP_{\text{rad}}}{dr}\right| = \frac{\kappa \rho F}{4\pi c r^2} < \left|\frac{dP}{dr}\right| = \frac{G \rho m}{r^2}, \tag{8.62}$$

or

$$\kappa F < 4\pi c G m. \tag{8.63}$$

When Eq. (8.63) is satisfied, then the star is said to be in radiative equilibrium. This condition can be violated by large fluxes due to intense nuclear burning. Near the center of the star, the core nuclear burning rate, q_c, is related to the flux by:

$$\frac{F}{m} \to q_c \quad \text{as} \quad m \to 0. \tag{8.64}$$

Therefore, a star has a radiative core as long as

$$q_c < \frac{4\pi cG}{\kappa}. \tag{8.65}$$

Equation (8.63) can also be violated by very high opacity, such as that which is usually encountered in regions where the temperature is close to the ionization temperature for the gas in the star. Near the surface of the star, $F = L$, and so the surface is in radiative equilibrium if

$$L < \frac{4\pi cGM}{\kappa}. \tag{8.66}$$

If the luminosity increases beyond this threshold, then the surface layer feels an outward pressure that is greater than the gravitational attraction and so it is blown off of the star and becomes unbound. This threshold luminosity is known as the *Eddington luminosity*: $L_{Edd} = 4\pi cGM/\kappa$. The opacity is often expressed relative to the electron scattering opacity, which is a constant, so

$$L_{Edd} = 3.2 \times 10^4 \left(\frac{M}{M_\odot}\right) \left(\frac{\kappa_{es}}{\kappa}\right) L_\odot. \tag{8.67}$$

One consequence of the Eddington luminosity is that it requires the existence of an upper bound on the masses of main sequence stars when combined with the mass luminosity relation ($L \propto M^\nu$).

Problem 8.5: Using $\nu = 3.5$, determine the upper bound on the mass of a main sequence star.

Problems

8.1. Solve the Lane–Emden equation for $n = 1$. Calculate the mass of the star in terms of the central density ρ_c.

8.2. The solar value of the mean molecular weight is 0.61. Use this to determine β for the Eddington Standard Model of the sun.

8.3. Using the value of β found in Problem 8.2 and the solutions for the $n = 3$ polytrope, determine the central density and central temperature of the sun in the Eddington Standard Model.

8.4. Numerically solve the $n = 3$ polytrope equation using solar values to determine the radius at which the temperature is high enough for the p–p chain to be operating ($T_6 = 15$).

8.5. Using $\nu = 3.5$, determine the upper bound on the mass of a main sequence star.

Chapter 9
Stability

The stellar models developed in the last chapter were valid for equilibrium states. However, equilibrium is not the same as stability. For example, a pencil balanced on its point is also in equilibrium, but it is not stable against small perturbations. We now need to consider when our equilibrium solutions are also stable against small perturbations. In so doing, we will encounter a number of potential instabilities that may have a significant effect in the structure and evolution of stars.

9.1 Thermal Stability

Usually, when a star is in hydrostatic equilibrium, it is stable. This is because of the temperature dependence of the nuclear reaction rates and the equation of state. To better understand the instabilities that may arise, it is useful to first look at the conditions that bring about stability.

The total energy of a star in hydrostatic equilibrium is the sum of the internal energy and the gravitational potential energy, but these two are related through the virial theorem, so

$$3\int_0^M \frac{P}{\rho} dm = -\Omega. \tag{9.1}$$

If the gas can also be described by the ideal gas law and the star has non-negligible radiation pressure, then

$$\frac{P}{\rho} = \frac{P_{\text{gas}}}{\rho} + \frac{P_{\text{rad}}}{\rho} = \frac{\mathscr{R}}{\mu}T + \frac{aT^4}{3\rho} = \frac{2}{3}u_{\text{gas}} + \frac{1}{3}u_{\text{rad}}. \tag{9.2}$$

Applying this to the virial theorem, we have

$$U_{\text{gas}} = -\frac{1}{2}(\Omega + U_{\text{rad}}), \tag{9.3}$$

M. Benacquista, *An Introduction to the Evolution of Single and Binary Stars*,
Undergraduate Lecture Notes in Physics, DOI 10.1007/978-1-4419-9991-7_9,
© Springer Science+Business Media New York 2013

so

$$E = U_{\text{gas}} + U_{\text{rad}} + \Omega = \frac{1}{2}(\Omega + U_{\text{rad}}) = -U_{\text{gas}}. \tag{9.4}$$

We also have
$$\dot{E} = L_{\text{nuc}} - L, \tag{9.5}$$

which simply states that if the power generated in the star through nuclear fusion is not equal to the power radiated from the surface, then the total energy will change. Combining Eqs. (9.4) and (9.5) leads to

$$L_{\text{nuc}} - L = -\dot{U}_{\text{gas}}. \tag{9.6}$$

In equilibrium, $\dot{E} = 0$ and $L_{\text{nuc}} = L$. We have seen in Chap. 7 that the rate of heat generated by nuclear reactions is strongly dependent upon the temperature in the core. Let us now consider the consequences of a small imbalance between the luminosity and the total rate of nuclear energy generation in the core. Equation (9.6) indicates that the internal energy of the star will drop if $L_{\text{nuc}} > L$ and it will increase if $L_{\text{nuc}} < L$.

If there is an increase in the nuclear generation rate or a decrease in the stellar luminosity, then $L_{\text{nuc}} > L$. For the star to be thermally stable, then this change must set off a chain of events that restores the balance between L_{nuc} and L by either lowering L_{nuc} or increasing L (or both). From Eq. (9.6), we see that when $L_{\text{nuc}} > L$, then $\dot{U}_{\text{gas}} < 0$. This implies that the internal energy of the gas will decrease, leading to a decrease in the temperature. A decrease in the temperature leads to a decrease in the internal energy of the radiation, so $\dot{U}_{\text{rad}} < 0$. From Eq. (9.5), we see that the total energy of the star must increase, so $\dot{E} > 0$, but the energy equation [Eq. (9.4)] requires that

$$\dot{\Omega} = \dot{E} - \dot{U}, \tag{9.7}$$

and so the gravitational potential energy of the star must increase. This requires that the star expand. If the star expands, then the average density must decrease. Consequently, both ρ and T decrease if $L_{\text{nuc}} - L$ increases. Since the nuclear energy generation rate is

$$q = q_0 \rho^\alpha T^\beta, \tag{9.8}$$

then q must also decrease. Finally, since

$$L_{\text{nuc}} = \int_0^M q \, dm, \tag{9.9}$$

the nuclear energy generation decreases and equilibrium is restored. A similar argument can be used in the case where $L_{\text{nuc}} < L$, leading to a contraction of the star and an increase in density and temperature. Thus, any small perturbations in the nuclear generation rate or the luminosity will result in a change in the nuclear generation rate that restores equilibrium.

9.2 Thermal Instability

The crucial assumption that was made in the previous section was that the gas followed the ideal gas law and so $P_{\text{gas}} \propto \rho^a T^b$. The chain of events described in the preceding section can also be explained as follows. An increase in L_{nuc} leads to an increase in temperature, which causes an increase in pressure (due to the ideal gas law), which causes the gas to expand, which does work on the surrounding star, causing the gas to expand and cool, lowering the temperature and returning L_{nuc} to equilibrium.

If the gas is degenerate, then $P_{\text{gas}} \sim P_e = K\rho^{5/3}$ (or $\rho^{4/3}$ if the gas is relativistic) and P_{gas} is independent of temperature. In this case an increase in L_{nuc} will not result in an increase in pressure, even though the temperature increases. To get a better understanding of this effect, we consider a homologous change in the star. By this, we mean that in time dt, a shell of radius r is displaced to $r + dr = r(1 + x dt)$, where x is a constant, so that all layers of the star increase in size with the same proportion. This implies that

$$\frac{\dot{r}}{r} = \frac{\partial \ln r}{\partial t} = x. \tag{9.10}$$

Since x is a constant throughout the star,

$$\frac{\partial}{\partial m}\left(\frac{\partial \ln r}{\partial t}\right) = 0, \tag{9.11}$$

and so

$$\frac{\partial}{\partial t}\left(\frac{\partial \ln r}{\partial m}\right) = \frac{\partial}{\partial t}\left(\frac{1}{r}\frac{\partial r}{\partial m}\right) = \frac{\partial}{\partial t}\left(\frac{1}{4\pi r^3 \rho}\right) = \frac{1}{4\pi r^3 \rho}\left(-3\frac{\dot{r}}{r} - \frac{\dot{\rho}}{\rho}\right) = 0, \tag{9.12}$$

or

$$\frac{\dot{\rho}}{\rho} = -3\frac{\dot{r}}{r}. \tag{9.13}$$

From the equation of hydrostatic equilibrium,

$$P = \int_0^M \frac{Gm}{4\pi r^4} dm, \tag{9.14}$$

we obtain

$$\dot{P} = \int_0^M \frac{\partial}{\partial t}\left(\frac{1}{r^4}\right)\frac{Gm}{4\pi} dm = \int_0^M -\frac{\dot{r}}{r}\frac{Gm}{4\pi r^4} dm = -4\frac{\dot{r}}{r}P. \tag{9.15}$$

We combine Eqs. (9.13) and (9.15) to get a relation between P and ρ:

$$\frac{\dot{P}}{P} = \frac{4}{3}\frac{\dot{\rho}}{\rho}, \tag{9.16}$$

which implies

$$\frac{dP}{P} = \frac{4}{3}\frac{d\rho}{\rho}. \tag{9.17}$$

We can look at a variety of equations of state by assuming a generic form

$$P = C\rho^a T^b, \tag{9.18}$$

where a and b are positive constants; then

$$\frac{dP}{P} = a\frac{d\rho}{\rho} + b\frac{dT}{T}, \tag{9.19}$$

and so:

$$\left(\frac{4}{3} - a\right)\frac{d\rho}{\rho} = b\frac{dT}{T}. \tag{9.20}$$

As long as $a < 4/3$, then an increase in density (due to contraction) leads to an increase in temperature. This is the case for ideal gases where $a = b = 1$. However, for degenerate gases, which have $a \geq 4/3$ and $b \ll 1$, the pressure is determined by $\rho_{electron}$ alone and an expansion would result in a slight increase in temperature. This is usually applied to the core of a star so that if the temperature increase leads to an increase in L_{nuc}, then the density will further decrease, leading to an increase in T. This is an unstable result and will continue until the temperature is so high and the density is so low such that the degeneracy is lifted.

9.3 Thin-Shell Instability

Another instance of thermal instability occurs when nuclear burning happens in a spherical shell. This can happen when nuclear fuel has been exhausted in the core but the temperatures are high enough outside of the core for fusion to continue in a layer surrounding the core. Consider a thin shell of mass Δm, temperature T, and density ρ, shown in Fig. 9.1. The shell lies between a fixed inner boundary r_0 (generally set by a core) and an outer boundary r, such that $\ell = (r - r_0) \ll r_0$. This is a thin shell. If the shell is in thermal equilibrium, then the rate of nuclear energy generation within the shell, $q(\Delta m)$, is equal to the net rate of heat flow out of the shell. If q increases, then the shell will expand and lift the layers above it. This, in turn, will result in a decrease in pressure. Depending upon the thickness of the shell and the equation of state of the gas, this can result in an instability. As we have seen, hydrostatic equilibrium requires

$$\frac{dP}{P} = -4\frac{dr}{r}. \tag{9.21}$$

9.3 Thin-Shell Instability

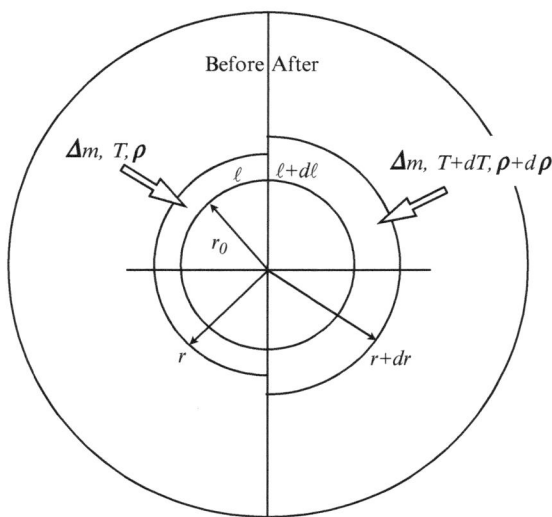

Fig. 9.1 Configuration of a nuclear burning thin shell. The configuration prior to expansion is on the *left* and the configuration of the shell after expansion is on the *right*. If the temperature increases after expansion, then the nuclear burning rate will increase, causing the expansion to continue

Since
$$\Delta m = 4\pi r_0^2 \ell \rho, \tag{9.22}$$

we have
$$\frac{d\rho}{\rho} = -\frac{d\ell}{\ell}. \tag{9.23}$$

Noting that $d\ell = dr$, this becomes
$$\frac{d\rho}{\rho} = -\frac{dr}{\ell} = -\frac{dr}{r}\frac{r}{\ell}, \tag{9.24}$$

and
$$\frac{dP}{P} = 4\frac{\ell}{r}\frac{d\rho}{\rho}. \tag{9.25}$$

Applying the generic equation of state from Eq. (9.18), we find
$$\left(4\frac{\ell}{r} - a\right)\frac{d\rho}{\rho} = b\frac{dT}{T}. \tag{9.26}$$

In this case, thermal stability requires that $a < 4\ell/r$. As can be seen, a sufficiently thin shell can always violate this condition for any value of a, at which point an expanding shell will result in an increase in temperature, which will increase the nuclear burning rate. The expansion will continue until ℓ becomes large enough for the stability condition to be met.

> **Problem 9.1:** A shell of hydrogen is burning around a helium core of radius r. If the gas is described by an ideal gas, what is the minimum outer radius of this shell if the hydrogen burning is stable? What is the minimum outer radius if the gas is a nonrelativistic degenerate gas?

Finally, we note that significant variations of F as a result of variations of T can alter this discussion as a change in F can also result in an increase in the temperature in a shell of the star. In completely ionized gases, F is much less sensitive to temperature than q is. However, instabilities can also occur at temperatures near the ionization energies of the constituent gases. Fluctuations about stable thermal equilibrium points and unstable deviations from equilibrium occur on thermal timescales. Thus, these instabilities usually occur over millions of years.

9.4 Dynamical Instabilities

Dynamical instabilities result in fluctuations (or catastrophic collapses or expansions) on dynamical timescales. The following discussion is a simplified example of how dynamical perturbations can lead to instabilities. We will consider perturbations in the pressure and density of a shell at $r(m)$. If the system is in equilibrium, then the pressure induced by the weight of the gas above the shell is equal to the internal pressure of the gas in the shell. For this configuration to be stable, any compression of the shell should result in an excess of pressure within the shell, exerting a restoring force that returns the shell to equilibrium.

The pressure is set by the equation of hydrostatic equilibrium:

$$P_h = \int_m^M \frac{Gm}{4\pi r^4} dm, \tag{9.27}$$

where we have assumed $P(M) = 0$ and we have retained the subscript h to indicate that this is the hydrostatic pressure needed to support the outer layer of the star. The density of the star at $r(m)$ is given by the continuity equation:

$$\rho = \frac{1}{4\pi r^2} \frac{dm}{dr}. \tag{9.28}$$

As before, we consider a homologous perturbation that results in a contraction:

$$r' = r(1 - \varepsilon), \tag{9.29}$$

where $\varepsilon \ll 1$. As a result of this contraction, the new density is

$$\rho' = \frac{1}{4\pi r'^2} \frac{dm}{dr'} = \frac{1}{4\pi r^2 (1-\varepsilon)^2} \frac{dm}{dr} \frac{dr}{dr'} = \frac{\rho}{(1-\varepsilon)^3} \simeq \rho(1+3\varepsilon). \tag{9.30}$$

9.4 Dynamical Instabilities

Now, if we are looking at dynamical perturbations, the timescale is short enough that we can assume the contraction is adiabatic, so

$$P'_{\text{gas}} = P\left(\frac{\rho'}{\rho}\right)^{\gamma_a} = P(1+3\varepsilon)^{\gamma_a} \simeq P(1+3\gamma_a\varepsilon). \tag{9.31}$$

The perturbation also changes the required hydrostatic pressure by

$$P'_h = \int_m^M \frac{Gm\,dm}{4\pi r^4 (1-\varepsilon)^4} \simeq P_h(1+4\varepsilon). \tag{9.32}$$

If the star is to remain stable, then the new gas pressure [Eq. (9.31)] must be greater than the new hydrostatic pressure [Eq. (9.32)] in order to restore equilibrium. Since $P_{\text{gas}} = P_h$ prior to the contraction, this implies

$$P'_{\text{gas}} > P'_h \implies 3\gamma_a > 4 \implies \gamma_a > \frac{4}{3}. \tag{9.33}$$

If $\gamma_a > \frac{4}{3}$ throughout the star, then there is global stability. If the adiabatic exponent varies throughout the star, it can be shown that if

$$\int_0^M \left(\gamma_a - \frac{4}{3}\right) \frac{P}{\rho} dm < 0, \tag{9.34}$$

then the star will be unstable. In other words, we weight the adiabatic exponent by the ratio of P to ρ, so that if $\gamma_a < \frac{4}{3}$ in the core of the star where P/ρ is large, it will be unstable even if $\gamma_a > \frac{4}{3}$ in the atmosphere. On the other hand, if $\gamma_a < \frac{4}{3}$ in the atmosphere but $\gamma_a > \frac{4}{3}$ in the core, the star can still be marginally stable.

Stars may become dynamically unstable near the end of their lives. If the core of a star reaches the Chandrasekhar mass and the gas becomes relativistic and degenerate, then $\gamma_a \to \frac{4}{3}$ and the star is at an inflection point. Any perturbation will lead to collapse that proceeds on a dynamical timescale. This is generally the fate of massive stars. If the pressure of a star becomes dominated by radiation pressure, then $\beta \to 0$. In this case, $\gamma_a \to \frac{4}{3}$. Then, the virial theorem gives $-\Omega = U_{\text{rad}}$ and so $E = \Omega + U_{\text{rad}} = 0$ and the star becomes unbound, in agreement with the discussion of the Eddington luminosity in Sect. 8.4.

When the temperature of the gas in a star is just at the threshold of ionization, the adiabatic exponent can drop below $\frac{4}{3}$ and lead to dynamical instability. In this case, if the number density of ions/electrons is low, then ionization is more likely to occur than recombination. If the number density is high, then recombination is more likely. Returning to Sect. 5.5 and Eq. (5.66) which gives the adiabatic exponent for a mixture of ionized and neutral gas and applying $kT \sim \xi$, we have

$$\gamma_a = \frac{20 + 49x(1-x)}{12 + 31x(1-x)}. \tag{9.35}$$

This gives $\gamma_a < \frac{4}{3}$ for $0.18 \leq x \leq 0.82$. Thus partially ionized gases near the ionization temperature can be unstable to dynamical perturbations.

If the star is dynamically stable, it will oscillate about its equilibrium state with a characteristic frequency. We can determine this frequency using the dynamical equation and considering Eq. (9.29) as a perturbation with a time-dependent ε. Starting with an equilibrium radius r_0 and pressure P_0, the dynamical equation gives

$$0 = -\frac{Gmdm}{r_0^2} - 4\pi r_0^2 dP_0, \qquad (9.36)$$

since $\ddot{r} = 0$ at equilibrium. Using $r = r_0(1-\varepsilon)$, the dynamical equation gives

$$\ddot{r}dm = -r_0 dm\ddot{\varepsilon} = -\frac{Gmdm}{r_0}(1-\varepsilon)^{-2} - 4\pi r_0^2(1-\varepsilon)^2 dP. \qquad (9.37)$$

Continuing with the assumption that $\varepsilon \ll 1$, we keep only terms of order ε in the dynamical equation and use Eq. (9.31) to find $dP = dP_0(1+3\gamma_a\varepsilon)$, so that

$$-r_0 dm\ddot{\varepsilon} = -\frac{Gmdm}{r_0^2}(1+2\varepsilon) - 4\pi r_0^2 dP_0(1-2\varepsilon+3\gamma_a\varepsilon). \qquad (9.38)$$

Substituting in for $4\pi r_0^2 dP_0$ from Eq. (9.36), we find

$$\ddot{\varepsilon} = -\frac{Gm}{r_0^3}(3\gamma_a - 4)\varepsilon. \qquad (9.39)$$

This is the equation for a simple harmonic oscillator with an oscillation frequency of

$$\omega = \sqrt{(3\gamma_a - 4)\frac{Gm}{r_0^3}}. \qquad (9.40)$$

If $\gamma_a < 4/3$, then the frequency is imaginary and ε grows exponentially, so the perturbation is unstable. If $\gamma_a > 4/3$, then the perturbation oscillates about the equilibrium state and the system is stable.

Problem 9.2: For an ideal gas, what is the stable oscillation frequency for $m = 1\,M_\odot$ and $r_0 = 1\,R_\odot$?

9.5 Convection

The dynamical stability discussed so far has been for shells of matter. Another instability involves individual packets of gas within a shell. This is convection. In convection, a packet of gas at slightly higher temperature and lower density than the

9.5 Convection

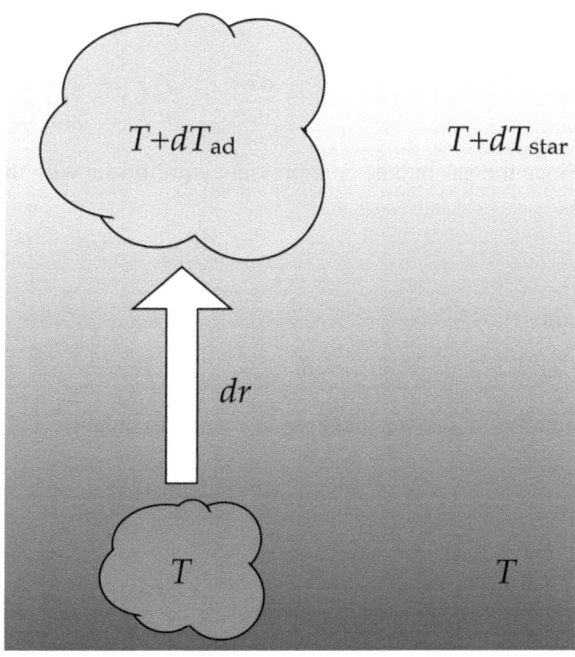

Fig. 9.2 The configuration for analyzing the criterion for convection. Note that since the temperature decreases toward the surface of a star, the temperature gradient in a star is negative. Therefore dT_ad and dT_star are both negative

surrounding gas will continue to rise until it is no longer hotter or less dense than its surroundings. In convection, the packet of gas is always assumed to have the same pressure as its surrounding gas.

When determining whether convection can be sustained, we need to look at the conditions that allow a rising bubble of gas to remain hotter than its surroundings. We consider a bubble of gas that rises and expands adiabatically as shown in Fig. 9.2. From the ideal gas law $P = (\mathscr{R}/\mu)\rho T$, we have

$$\frac{dP}{dr} = -\frac{P}{\mu}\frac{d\mu}{dr} + \frac{P}{\rho}\frac{d\rho}{dr} + \frac{P}{T}\frac{dT}{dr}, \tag{9.41}$$

and if we assume that the composition doesn't change, then $\mu = $ constant, and

$$\frac{dP}{dr} = \frac{P}{\rho}\frac{d\rho}{dr} + \frac{P}{T}\frac{dT}{dr}. \tag{9.42}$$

Since the expansion is adiabatic, we have $P = K\rho^{\gamma_a}$, and so

$$\frac{dP}{dr} = \gamma_a \frac{P}{\rho}\frac{d\rho}{dr}. \tag{9.43}$$

Combining these two equations, we can obtain the temperature gradient of the rising gas bubble:

$$\left.\frac{dT}{dr}\right|_{ad} = \frac{T}{P}\left(1 - \frac{1}{\gamma_a}\right)\frac{dP}{dr}. \tag{9.44}$$

Since the gas bubble is in pressure equilibrium with the surrounding gas,

$$\frac{dP}{dr} = -\rho\frac{Gm}{r^2}, \tag{9.45}$$

and

$$\left.\frac{dT}{dr}\right|_{ad} = \frac{T}{P}\left(1 - \frac{1}{\gamma_a}\right)\left(-\rho\frac{Gm}{r^2}\right). \tag{9.46}$$

Finally, we use the ideal gas law again to obtain

$$\frac{T}{P} = \frac{\mu}{\mathcal{R}\rho}, \tag{9.47}$$

and

$$\left.\frac{dT}{dr}\right|_{ad} = -\frac{Gm\mu}{\mathcal{R}r^2}\left(1 - \frac{1}{\gamma_a}\right). \tag{9.48}$$

This is the change in temperature of the bubble as it rises.

If the actual temperature gradient in the star is steeper than the adiabatic temperature gradient, then the bubble will remain hotter (and therefore less dense) than its surroundings and continue to rise. Thus the condition for convection is

$$\left|\frac{dT}{dr}\right|_{star} > \left|\frac{dT}{dr}\right|_{ad} = \frac{GM\mu}{\mathcal{R}r^2}\left(1 - \frac{1}{\gamma}\right). \tag{9.49}$$

Once this condition has been met, convection will be the dominant energy transport mechanism and the equation

$$\frac{dT}{dr} = -\frac{3}{4ac}\frac{\kappa\rho}{T^3}\frac{F}{4\pi r^2} \tag{9.50}$$

will no longer be valid—unless F is replaced by a function that includes energy transport by convection.

Problem 9.3 Starting with Eq. (9.44), show that the condition for convection can also be written

$$\frac{d\ln P}{d\ln T} < \frac{\gamma_a}{\gamma_a - 1}.$$

9.6 Mixing Length Theory

The details of energy transport by convection are not well modeled, but an empirical model called the "mixing length theory" can provide an approximation of the convective flux, F_c. In this model, we assume that the gas bubble will eventually transfer its excess heat to the surrounding stellar material after traveling a characteristic distance ℓ, called the mixing length. The mixing length is expressed as $\ell = \alpha H_p$, where $H_p = -(\partial \ln P/\partial r)^{-1}$ is the scale height of pressure variation in the star and α is some constant to be determined empirically. The convective flux will then be the rate at which this heat is transferred during the bubble's rise. Since the process of convection is a constant pressure process, the amount of heat that flows from the bubble to the surroundings over the mixing length can be written:

$$\delta Q = \rho C_p \delta T_f, \tag{9.51}$$

where C_p is the specific heat at constant pressure and δT_f is the temperature difference between the bubble and its surroundings if no heat flow had occurred:

$$\delta T_f = \left(\left.\frac{dT}{dr}\right|_{ad} - \left.\frac{dT}{dr}\right|_{star}\right)\ell = \delta\left(\frac{\partial T}{\partial r}\right). \tag{9.52}$$

Thus, in order to obtain an expression for the convective flux, we need to determine the rate at which the bubbles transfer this heat. If we assume the bubbles travel at an average speed $\bar{v}_c$, then the rate of heat transfer over the mixing length is

$$F_c = \delta Q \bar{v}_c = C_p \delta T_f \rho \bar{v}_c. \tag{9.53}$$

Now, we need to estimate $\bar{v}_c$. We do this by assuming that the kinetic energy gained by the bubble is equal to the work done on the bubble by the buoyant force and that $\bar{v}_c$ is proportional to the average speed found from the kinetic energy. The net force is

$$f = -g\delta\rho, \tag{9.54}$$

where $\delta\rho$ is the density difference between the bubble and its surroundings and g is the local acceleration:

$$g = \frac{Gm}{r^2}. \tag{9.55}$$

The fluid is an ideal gas under constant pressure, so

$$\delta P = \frac{P}{\rho}\delta\rho + \frac{P}{T}\delta T = 0, \tag{9.56}$$

$$\delta\rho = -\frac{\rho}{T}\delta T, \tag{9.57}$$

$$f = \rho g \left(\frac{\delta T}{T}\right), \qquad (9.58)$$

where δT is the instantaneous temperature difference and varies from 0 at the beginning up to δT_f after the bubble has traveled ℓ. Thus, we will approximate the average force as

$$f = \frac{1}{2}\rho g \frac{\delta T_f}{T}. \qquad (9.59)$$

Thus, the work done by the buoyant force is $f\ell$.

The average kinetic energy is $\frac{1}{2}\rho \langle v^2 \rangle$, and we will express $\bar{v}_c$ as $\bar{v}_c^2 = \alpha' \langle v^2 \rangle$, where α' is another constant to be determined empirically. Therefore

$$\frac{1}{2}\rho g \ell \frac{\delta T_f}{T} = \frac{1}{2}\rho \frac{\bar{v}_c^2}{\alpha'}, \qquad (9.60)$$

so

$$\bar{v}_c = \left[\frac{\alpha' g}{T}\right]^{1/2} \left[\delta\left(\frac{\partial T}{\partial r}\right)\right]^{1/2} \ell. \qquad (9.61)$$

Finally, we have

$$F_c = \delta Q \bar{v}_c = C_p \left[\delta\left(\frac{\partial T}{\partial r}\right)\right]^{3/2} \left[\frac{\alpha' g}{T}\right]^{1/2} \rho \ell^2. \qquad (9.62)$$

The pressure scale height can be written

$$H_p = \frac{P}{\rho g} = \frac{\mathscr{R} T}{\mu g}, \qquad (9.63)$$

so

$$\ell = \alpha H_p = \alpha \frac{\mathscr{R} T}{\mu g} \qquad (9.64)$$

and

$$F_c = C_p \left[\delta\left(\frac{\partial T}{\partial r}\right)\right]^{3/2} \left[\frac{\alpha' g}{T}\right]^{1/2} \rho \alpha^2 \left[\frac{\mathscr{R}}{\mu}\right]^2 \left[\frac{T}{g}\right]^2$$

$$= \alpha^2 \alpha'^{1/2} \rho C_p \left[\frac{\mathscr{R}}{\mu}\right]^2 \left[\frac{T}{g}\right]^{3/2} \left[\delta\left(\frac{\partial T}{\partial r}\right)\right]^{3/2}. \qquad (9.65)$$

This theory requires two "fudge factors" (α and α') that are of order unity, although there is a stronger dependence on α. Additionally, there is a fairly strong dependence on the differences in temperature gradients. Since a larger value of F_c will tend to reduce this difference, the net result is that convection nearly always results in a temperature gradient that is equal to the adiabatic temperature gradient. Consequently, F_c is chosen in order to maintain the adiabatic temperature gradient when the star is unstable with respect to convection.

Problems

9.1. A shell of hydrogen is burning around a helium core of radius r. If the gas is described by an ideal gas, what is the minimum outer radius of this shell if the hydrogen burning is stable? What is the minimum outer radius if the gas is a nonrelativistic degenerate gas?

9.2. For an ideal gas, what is the stable oscillation frequency for $m = 1\,M_\odot$ and $r_0 = 1\,R_\odot$?

9.3. Starting with Eq. (9.44), show that the condition for convection can also be written

$$\frac{d\ln P}{d\ln T} < \frac{\gamma_a}{\gamma_a - 1}.$$

Chapter 10
Stellar Birth

We are now in a position to begin describing the evolution and structure of stars. We have the dynamical equations that were developed in Chap. 4. We have the descriptions of the auxiliary variables, P, F, and q, that were developed in Chaps. 5–7. Finally, we have an understanding of the appropriate instabilities developed in Chap. 9. With all of these tools in hand, we will start with the birth of stars from clouds of cold interstellar gas and dust.

10.1 The Jeans Criteria

Most stars in the Galaxy are thought to have formed in clusters through the collapse of cold interstellar clouds. These clouds consist of molecular hydrogen as well as dust and other molecules. They have temperatures in the range of 10–100 K and number densities around 10^8 m^{-3}. The total masses are about 10^6 M$_\odot$ and typical sizes are about 30 pc (100 lyr) across. Typically, these clouds are in hydrostatic equilibrium until some perturbation drives them out of equilibrium. By this we mean that the perturbation must be large enough to push the cloud outside of the stability regime so that the cloud starts to collapse. When collapse begins, the cloud will fragment into separate collapsing regions whose size and mass can be estimated using the equation for radial evolution for a spherically symmetric cloud:

$$\frac{d^2 r}{dt^2} = -\frac{Gm(r)}{r^2} - \frac{1}{\rho}\frac{\partial P}{\partial r} = -\frac{Gm}{r^2} - 4\pi r^2 \frac{\partial P}{\partial m}. \tag{10.1}$$

If the pressure term dominates, then the cloud will expand on a timescale governed by the speed of sound in the gas, c_s. For a cloud of mass m, density ρ, and temperature T, this timescale is

$$t_s \simeq \frac{R}{c_s} \propto \frac{R \rho^{1/2}}{P^{1/2}} \propto m^{1/3} \rho^{-1/3} T^{-1/2}. \tag{10.2}$$

On the other hand, if the self-gravity term dominates, then the cloud will collapse at a rate governed by free fall:

$$t_{\text{ff}} \simeq \left(\frac{Gm}{r^3}\right)^{-1/2} \propto (G\rho)^{-1/2}. \tag{10.3}$$

The entire cloud does not collapse, only those regions with $t_{\text{ff}} \ll t_s$ will do so. Thus, in regions where $t_{\text{ff}} \ll t_s$, deviations from equilibrium will tend to result in collapse. This provides us with the means to estimate the mass of a collapsing cloud in terms of the density and temperature.

We can obtain an expression of the typical mass of a collapsing cloud by considering the virial theorem, written as

$$\int_{P_c}^{P_s} V\,dP = -\frac{1}{3}\int_0^{M_s} \frac{Gm\,dm}{r}, \tag{10.4}$$

where the subscript "s" indicates the value on the surface of the collapsing part of the cloud and P_c is the pressure at the center of the collapsing part. Using integration by parts, we have

$$\int_{P_c}^{P_s} V\,dP = [PV]\Big|_{P_c}^{P_s} - \int_0^{V_s} P\,dV. \tag{10.5}$$

Because $V_c = 0$, $[PV]\Big|_{P_c}^{P_s} = P_s V_s$ and

$$\int_0^{V_s} P\,dV = P_s V_s + \frac{1}{3}\int_0^{M_s} \frac{Gm\,dm}{r}$$

$$= P_s V_s + \frac{\alpha}{3}\frac{GM_s^2}{R_s}, \tag{10.6}$$

where α depends upon the internal structure of the cloud. For the conditions in molecular clouds, the gas can be assumed to be isothermal, since the optical depth is nearly zero at these temperatures and densities. Therefore,

$$\int_0^{V_s} P\,dV = \frac{\mathscr{R}}{\mu}T\int_0^{V_s}\rho\,dV = \frac{\mathscr{R}}{\mu}TM_s, \tag{10.7}$$

or

$$\frac{\mathscr{R}}{\mu}TM_s = P_s V_s + \frac{\alpha}{3}\frac{GM_s^2}{R_s}. \tag{10.8}$$

Since V_s is clearly nonnegative and P_s is the external pressure pushing on the cloud (so $P_s \geq 0$), then equilibrium cannot occur if

$$\frac{\mathscr{R}}{\mu}TM_s < \frac{\alpha}{3}\frac{GM_s^2}{R_s}. \tag{10.9}$$

10.1 The Jeans Criteria

Defining the mean density to be $\rho_0 = M_s/(\frac{4}{3}\pi R_s^3)$, we have

$$R_s = \left(\frac{3M_s}{4\pi\rho_0}\right)^{1/3} \tag{10.10}$$

and

$$\frac{\mathscr{R}}{\mu}T < \frac{\alpha}{3}GM_s\left(\frac{4\pi\rho_0}{3M_s}\right)^{1/3} = \frac{\alpha}{3}G\left(\frac{4\pi\rho_0}{3}\right)^{1/3}M_s^{2/3}. \tag{10.11}$$

Solving for the mass, we have

$$M_s > \left[\frac{\mathscr{R}}{\mu}T\frac{3}{\alpha G}\left(\frac{3}{4\pi\rho_0}\right)^{1/3}\right]^{3/2} = M_J, \tag{10.12}$$

where the *Jeans mass* is defined as

$$M_J = \left[\frac{\mathscr{R}T}{\mu G}\right]^{3/2}\left[\frac{3}{4\pi}\right]^{1/2}\left[\frac{3}{\alpha}\right]^{3/2}\rho_0^{-1/2}. \tag{10.13}$$

If we assume constant density, then $\alpha = 3/5$, and

$$M_J = \left[\frac{5\mathscr{R}T}{\mu G}\right]^{3/2}\left[\frac{3}{4\pi\rho_0}\right]^{1/2}. \tag{10.14}$$

For typical values found in interstellar clouds, this is

$$M_J = 1.2 \times 10^5\,M_\odot\left(\frac{T}{100\,\text{K}}\right)^{3/2}\left(\frac{\rho_0}{10^{-21}\,\text{kg/m}^3}\right)^{-1/2}\mu^{-3/2}. \tag{10.15}$$

Problem 10.1: The Jeans length (R_J) is defined to be the minimum radius necessary to collapse a cloud of density ρ_0.

(a) Use the expression of the Jeans mass to obtain the following one for the Jeans length:

$$R_J = \sqrt{\frac{15\mathscr{R}T}{4\pi\mu G\rho_0}}.$$

(b) For a typical diffuse hydrogen cloud, $T = 50\,\text{K}$ and $n = 5 \times 10^8\,\text{m}^{-3}$. If we assume that the cloud is entirely composed of H I, $\rho_0 = m_H n = 8.4 \times 10^{-19}\,\text{kg/m}^3$. Taking $\mu = 1$, determine R_J for this cloud.

In an interstellar gas cloud, only objects with $M \geq M_J$ can begin to collapse. Of course, most stars are considerably less massive than $10^5 \, M_\odot$. Clearly some other effects must eventually become important in order to reduce the mass of collapsing objects to typical stellar masses. Initially, the cloud is transparent to the blackbody radiation emitted by a cold cloud, so the collapse can be considered isothermal as any increase in temperature is immediately radiated away. However, the density will naturally increase and thus, since $M_J \propto T^{3/2} \rho^{-1/2}$, the Jeans mass will decrease. This means that smaller subregions of the collapsing cloud that are slightly overdense to start with can then begin to collapse. This process is known as *fragmentation*.

Fragmentation will continue until the temperature begins to rise. While the collapsing cloud is transparent, the collapse is isothermal and continues on a free-fall timescale. Once the density of the gas is high enough that the cloud becomes opaque and can no longer radiate away the energy gained through gravitation, then the collapse will become adiabatic. Generally at this point, the gas can be considered to be a monatomic ideal gas, and so the adiabatic collapse proceeds along $P \propto T^{5/2} \propto \rho^{5/3}$, so $T \propto \rho^{2/3}$ and $M_J \propto \rho \rho^{-1/2} = \rho^{1/2}$, and the fragmentation ceases since the Jeans mass now increases with increasing density. To estimate the lowest mass stars that can be formed by this process, we note that the rate of energy loss during isothermal collapse is

$$A \simeq \frac{E}{t_{\rm ff}} = \frac{GM^2}{R} (G\rho)^{1/2} = \left(\frac{3}{4\pi}\right)^{1/2} \frac{G^{3/2} M^{5/2}}{R^{5/2}}. \tag{10.16}$$

This rate cannot be larger than the blackbody rate for a cloud of the same temperature:

$$B = (4\pi R^2)(\sigma T^4) f, \tag{10.17}$$

where f is some fraction less than 1. During isothermal collapse, $B \gg A$. When $B \simeq A$, then the transition to adiabatic collapse occurs. Equating B and A gives

$$M^5 = \left[\frac{64\pi^3}{3}\right] \left[\frac{\sigma^2 f^2 T^8 R^9}{G^3}\right]. \tag{10.18}$$

Replacing R by $3M/4\pi\rho_0$ and applying Eq. (10.14) to eliminate ρ_0, we find

$$M_J = \left(\frac{3 \times 5^9}{64\pi^3}\right)^{1/4} \frac{1}{\sqrt{\sigma G^3}} \left(\frac{k}{\mu m_H}\right)^{9/4} f^{-1/2} T^{1/4}. \tag{10.19}$$

If $f = 1$, then we can set a lower limit on M_J of $M_J \geq 0.036$. More reasonable values of $f \sim 0.1$ and $T \sim 10^3 \, \rm K$ give $M_J \geq 0.36$.

10.2 Formation of a Protostar

In the preceding section, we made a number of simplifying assumptions in order to obtain an analytical description of the dynamically unstable and inherently nonlinear process of the collapse and fragmentation of an interstellar cloud to form stellar mass objects that will further collapse to form stars. After fragmentation, the process involves more of the dynamical evolution equations once the opacity, equation of state, molecular dissociation, and atomic ionization begin to come into play. Eventually, the central densities and temperatures will become high enough for nuclear processes to begin. A comprehensive treatment of star formation requires numerical solution to these coupled, nonlinear equations.

Nonetheless, certain gross features and processes can be understood from a more qualitative description of the process of the formation of a protostar. Here we will look at the evolution of a $1\,M_\odot$ fragment from isothermal collapse to nuclear ignition. For this example, we will take the initial temperature of the fragment to be $T_i = 50\,K$. From this information and the assumption that the fragment is initially in virial equilibrium, we can obtain the initial size and luminosity of the fragment. From virial equilibrium, we have

$$\frac{1}{2}Mv_{\text{rms}}^2 \simeq \frac{3}{5}\frac{GM^2}{R_i}, \tag{10.20}$$

and $v_{\text{rms}} = \sqrt{3kT_i/\mu m_H} = 1000\,\text{m/s}$ for $T_i = 50\,K$. Therefore,

$$R_i = \frac{2}{5}\frac{GM\mu m_H}{kT_i} \simeq 2\times 10^5\,R_\odot, \tag{10.21}$$

and the luminosity is

$$L_i = 4\pi\sigma R_i^2 T_i^4 \simeq 10^2\,L_\odot. \tag{10.22}$$

We can also obtain the particle number density:

$$n_i = \frac{3M}{4\pi\mu m_H R_i^3} \simeq 10^{14}\,\mu^{-1}\,\text{m}^{-3}. \tag{10.23}$$

If it were visible, the fragment would appear above and to the right of its final location on the main sequence on the H-R diagram.

The initial collapse will be isothermal, so the track would be a vertical line in the H-R diagram. Since the radius will decrease and $L \propto R^2 T_{\text{eff}}^4$, the luminosity drops. Although the temperature is constant, the pressure for an ideal gas will increase as the density increases. In particular, consider the radial evolution equation:

$$\ddot{r} = -\frac{Gm}{r^2} - 4\pi r^2\frac{\partial P}{\partial m} = -\frac{Gm}{r^2} - \frac{1}{\rho}\frac{\partial P}{\partial r}. \tag{10.24}$$

For an order of magnitude estimate, note that

$$\frac{1}{\rho}\frac{\partial P}{\partial r} \simeq \frac{1}{\rho_0}\frac{P_c}{R} \propto \frac{T}{R}, \qquad (10.25)$$

where P_c is the central pressure and

$$-\frac{Gm}{r^2} \simeq -\frac{GM}{R^2}. \qquad (10.26)$$

Therefore, for constant M and T, gravity dominates once the collapse begins and R shrinks. We can assume the pressure term is negligible during this first stage of collapse, so the timescale for the isothermal collapse is the free-fall timescale.

We can quickly estimate the free-fall timescale by considering the time it takes for a test mass to fall from R to 0 due to the gravity of a point mass M at the origin. This can be calculated from Kepler's third law, using an orbit with $e = 0$ and $2a = R$:

$$t_{\rm ff} = \frac{P_{\rm orb}}{2} = \sqrt{\frac{4\pi^2 a^3}{4GM}} = \sqrt{\frac{\pi^2 R^3}{8GM}} = \sqrt{\frac{3\pi}{32G\rho_0}}. \qquad (10.27)$$

For the fragment under discussion, $T \sim 50\,{\rm K}$, $n = 10^{14}\,{\rm m}^{-3}$, $\rho \simeq m_{\rm H} n \simeq 2 \times 10^{-13}\,{\rm kg/m}^3$, then

$$t_{\rm ff} \simeq 4,700\,{\rm years}. \qquad (10.28)$$

Note that $t_{\rm ff} \propto \rho^{-1/2}$, and so the central regions will be more dense, and their density will increase during the collapse. The optical depth is given by $\tau = \kappa \rho R$ and is proportional to density for the central regions of the star. At the beginning, the opacity is mainly due to dust, so $\kappa \simeq 10^{-3}\,{\rm m}^2/{\rm kg}$ and the density is $\rho \simeq 10^{-10}\,{\rm kg/m}^3$. Once the density becomes large enough, the star becomes opaque as the optical depth begins to exceed $\tau = 2/3$ at its center. At this point, the rate of collapse slows substantially in the core as the internal temperature and pressure begin to increase, while the photosphere begins to move out from the center. This leads to a central object called a *protostar* with freely falling gas around it. The temperature and luminosity of the protostar are set by the photosphere which is defined as the radius where $\tau = 2/3$. This radius is typically out where the temperature is still $\sim 50\,{\rm K}$.

Problem 10.2: Using the opacity for dust and taking the initial density to be $\rho \simeq 10^{-10}\,{\rm kg/m}^3$, determine the radius of the star when the optical depth becomes $2/3$ at the center. Assume constant density.

The protostar is in hydrostatic equilibrium, and the gas from the collapsing cloud continues to fall onto the surface of the protostar. A shock wave develops once the speed of the in-falling material exceeds the local speed of sound in the

10.2 Formation of a Protostar

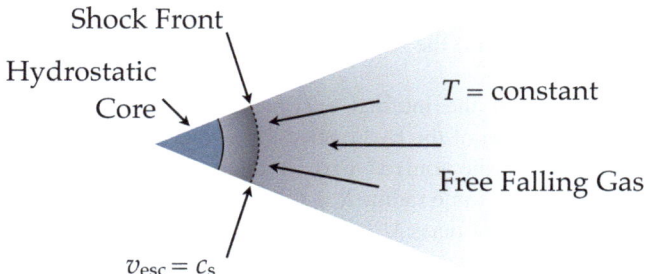

Fig. 10.1 A cartoon of the structure of the collapsing cloud with a hydrostatic protostar, surrounded by a shock front, embedded in a freely falling collapsing cloud

gas (Fig. 10.1). From the perspective of the gas, the surface of the protostar is crashing into it at supersonic speeds. Remembering that we are still working on the assumption of spherical symmetry, the accretion rate of material onto the shock front can be approximated by assuming that the mass of a shell ($m(r)$) falls onto the front in the free-fall time from radius r. This can be related to the escape velocity by

$$\dot{M} = \frac{m(r)}{t_{\text{ff}}(r)} \simeq \frac{v_{\text{esc}}^2 r/G}{r/v_{\text{esc}}} = \frac{v_{\text{esc}}^3(r)}{G}, \quad (10.29)$$

where $v_{\text{esc}}^2(r) = Gm(r)/r$. The shock front is found when the escape velocity equals the speed of sound, $v_{\text{esc}} = c_s$, so

$$\dot{M} = \frac{c_s^3}{G} = 2 \times 10^{-6} \left(\frac{T}{10\,\text{K}}\right)^{3/2} M_\odot\,\text{year}^{-1}. \quad (10.30)$$

Thus, a protostar of mass M is built up in a time given by

$$t_* \simeq 5 \times 10^5 \left(\frac{M}{M_\odot}\right) \left(\frac{T}{10\,\text{K}}\right)^{-3/2} \text{year}. \quad (10.31)$$

The next stage in evolution occurs as the central temperature of the protostar reaches $\sim 2000\,\text{K}$. At this temperature the molecular hydrogen begins to dissociate. Similar to the discussion of ionization, dissociation absorbs energy and doubles the number of gas particles. This tends to decrease the adiabatic exponent. Since the adiabatic exponent for a diatomic molecule is $7/5 = 1.4$, a reduction in the adiabatic exponent results in a dynamical instability where $\gamma_a < 4/3$. The result is that the core becomes unstable and begins to collapse again. Qualitatively, the energy of collapse goes into dissociation rather than increasing the temperature. After dissociation is complete, the adiabatic exponent grows to $\gamma_a = 5/3$ for a monatomic gas and the collapse halts. Another shock wave develops in the envelope, but there is very little gas left in the envelope, and the growth of the mass of the star essentially stops. The collapse of the core now begins to cause the temperature to rise and eventually hydrogen and helium are ionized.

If we assume that all the internal energy increase from collapse goes into dissociation and ionization of the hydrogen and helium, then we can find an upper bound on the radius of the ionized part of the forming protostar. The energy available from collapse to some radius R is $GM^2/2R$. The energy required to ionize a component gas in a star of mass M is

$$E = M \left(\frac{X_i}{\mathscr{A}_i m_H} \right) \xi, \tag{10.32}$$

where ξ is the ionization energy per atom and $\mathscr{A}_i$ is the atomic mass of component species i, which has mass fraction X_i in the star. For dissociation, the equation is similar in form, but $\mathscr{A}$ is replaced by the molecular mass and ξ is now the dissociation energy. Thus, the total energy required to dissociate and ionize the hydrogen and helium in a star of mass M is

$$E = \frac{M}{4m_H} [X(4\xi_H + 2\xi_d) + Y\xi_{He}], \tag{10.33}$$

where the dissociation energy of molecular hydrogen is $\xi_d = 4.48\,\text{eV}$, and the ionization energies of hydrogen and helium are $\xi_H = 13.6\,\text{eV}$ and $\xi_{He} = 78.98\,\text{eV}$, respectively. If we now assume that the metal content of the collapsing cloud is negligible, we can set $Y = 1 - X$ and then

$$E = \frac{M\xi_{He}}{4m_H} \left[1 + \left(\frac{4\xi_H + 2\xi_d}{\xi_{He}} - 1 \right) X \right] = (3 \times 10^{39}\,\text{J}) (1 - 0.2X) \frac{M}{M_\odot}. \tag{10.34}$$

Equating this energy to the energy released from collapse to R allows us to solve for the final radius:

$$\frac{R}{R_\odot} = \frac{GM_\odot^2}{2R_\odot (3.8 \times 10^{39}\,\text{J})(1 - 0.2X)} \frac{M}{M_\odot} \simeq \frac{50}{1 - 0.2X} \frac{M}{M_\odot}. \tag{10.35}$$

For a typical primordial composition of $X = 0.75$, the size of a $1\,M_\odot$ protostar is approximately $60\,R_\odot$. The central temperature of the protostar is

$$T \simeq \frac{GMm_H}{\kappa R} \sim 10^5\,\text{K}. \tag{10.36}$$

The photosphere is still governed by the dust opacity and so the effective temperature still remains constant at $T_{\text{eff}} \sim 50\,\text{K}$. Since

$$L \propto T_{\text{eff}}^4 R^2, \tag{10.37}$$

10.2 Formation of a Protostar

so

$$\frac{L_i}{L_f} = \frac{R_i^2}{R_f^2} = \left(\frac{10^5}{60}\right)^2 = 3 \times 10^6 \qquad (10.38)$$

and the final luminosity is $L_f \simeq 10^{-4} L_\odot$.

Problem 10.3: The typical metal composition of a star with solar metallicity is described by $X = 0.68$, $Y = 0.3$, and $Z = 0.02$. Using these values, determine the radius and central temperature of the star after ionization of hydrogen and helium.

As the outer edge of the star moves inward, the effective temperature now begins to rise. When $T_{\text{eff}} \sim 10^3$ K, the dust vaporizes and the opacity drops substantially and so the photosphere approaches the surface of the hydrostatic core. At this point a strong temperature gradient develops and convection becomes the dominant energy transport mechanism. This part of the contraction is very poorly modeled, so the exact process is uncertain, but the end result is that the effective temperature is increased to ~ 4000 K and the radius is $\sim 60 R_\odot$. Because the main process at work here is convection, this stage occurs quickly at ~ 300 days. A qualitative way of describing this process is to consider that the luminosity is produced by the release of gravitational energy:

$$E \simeq \frac{GM^2}{R} \qquad (10.39)$$

in a timescale of

$$t \simeq \left(\frac{GM}{R^3}\right)^{-1/2}, \qquad (10.40)$$

so

$$L \sim \frac{E}{t} \propto R^{-5/2}. \qquad (10.41)$$

Now, the effective temperature is $T_{\text{eff}} \propto \left(L/R^2\right)^{1/4} \propto R^{-9/8}$, so

$$L \propto T_{\text{eff}}^{20/9}. \qquad (10.42)$$

This change in the effective temperature continues until the dense core is essentially complete. Then, the luminosity is governed by the release of energy from matter accreted from the gas cloud, so

$$L \propto GM\dot{M}/R. \qquad (10.43)$$

Initially, L increases as M increases and R decreases, but as M and R stabilize, L is essentially constant and fixed by $\dot{M}$.

10.3 Contraction to Main Sequence

At the end of the formation of the protostar, the cloud of gas has become a fully convective ball of radius $R \sim 50 R_\odot$, $T_{\text{eff}} \sim 4{,}000\,\text{K}$, and $L \sim 100 L_\odot$. This ball will continue to collapse since the only source of energy for the luminosity is gravitational energy. Thus, the temperature of the star will continue to increase. Eventually, the temperature of the outer atmosphere is great enough to partially ionize the heavier elements. The electrons released by this process can form H^- ions. The opacity law for H^- is

$$\kappa_{H^-} \propto \rho^{1/2} T^9. \tag{10.44}$$

For a protostar that is fully convective, we can obtain a relationship between T_{eff}, M, and L for an opacity law of the form $\kappa = \kappa_0 \rho^n T^{-s}$. The thin radiative atmosphere of the protostar governs the rate at which energy can be emitted from the protostar, while the convective bulk of the protostar governs the rate at which energy is brought up to the surface. At the boundary between the atmosphere and the bulk of the protostar, we can equate the pressure calculated down from the atmosphere to that calculated up from the bulk, which will relate the effective temperature to the bulk properties of the protostar.

Near the surface of the star the structure equations read

$$\frac{dP}{dm} = -\frac{GM}{4\pi R^4}, \tag{10.45}$$

$$\frac{dT}{dm} = -\frac{3}{4ac}\frac{\kappa}{T^3}\frac{L}{(4\pi R^2)^2}. \tag{10.46}$$

Combining these yields

$$\frac{dP}{dT} = \frac{16\pi ac GM T^3}{3\kappa L}. \tag{10.47}$$

Using the ideal gas law, we can write the density in terms of the pressure and temperature to find

$$\rho = \frac{P}{T}\frac{\mu}{\mathscr{R}}, \tag{10.48}$$

so using $\kappa = \kappa_0 \rho^n T^{-s}$ gives

$$\frac{dP}{dT} = \frac{16\pi ac GM}{3\kappa_0 L}\left(\frac{\mathscr{R}}{\mu}\right)^n P^{-n} T^{n+s+3}. \tag{10.49}$$

Integrating this equation gives us the relation for the pressure near the surface of the star:

$$P_s^{n+1} \propto \frac{M}{L} T^{n+s+4}. \tag{10.50}$$

10.3 Contraction to Main Sequence

Within the convective bulk of the protostar, we know that the pressure and temperature are related by the condition for convection, so

$$\frac{dP}{dT} = \frac{\gamma_a}{\gamma_a - 1}. \tag{10.51}$$

For a monatomic ideal gas, $\gamma_a = 5/3$ and therefore

$$P = KT^{5/2}, \tag{10.52}$$

where K is a constant that depends on the mass and radius of the protostar. We can determine this dependence by expressing Eq. (10.52) in terms of a dimensionless pressure and temperature. From the hydrostatic equation and dimensional considerations, we define a dimensionless pressure $p = P/P_0$, where

$$P_0 = \frac{GM^2}{R^4}. \tag{10.53}$$

Using the ideal gas law, we can define a dimensionless temperature $t = T/T_0$ by

$$P = pP_0 \propto \left(\frac{\mathscr{R}}{\mu}\right) \frac{3M}{4\pi R^3} T_0 t, \tag{10.54}$$

so

$$T_0 = P_0 \left(\frac{\mu}{\mathscr{R}}\right) \frac{R^3}{M} = \left(\frac{\mu}{\mathscr{R}}\right) \frac{GM}{R}. \tag{10.55}$$

Returning to Eq. (10.52), we have

$$pP_0 = KT_0^{5/2} t^{5/2}. \tag{10.56}$$

Therefore,

$$K = \frac{P_0}{T_0^{5/2}} \propto M^{-1/2} R^{-3/2}. \tag{10.57}$$

Finally, using the Stefan–Boltzmann law, we can write $R \propto L^{1/2} T^{-2}$, and the pressure relation for the convective bulk is

$$P_b \propto M^{-1/2} L^{-3/4} T^{11/2}. \tag{10.58}$$

Equating P_b and P_s at the photosphere gives us the following relation for T_{eff} in a fully convective photosphere:

$$T_{\text{eff}} \propto \left(\frac{M}{M_\odot}\right)^{7/51} \left(\frac{L}{L_\odot}\right)^{1/102} \text{K}. \tag{10.59}$$

This is a very weak dependence on L, so as the star contracts, the luminosity drops, but there is almost no change in $T_{\rm eff}$. As the star contracts, it follows an almost vertical line in the H-R diagram known as a Hayashi track. Stars will continue to move along Hayashi tracks until the temperature gradient favors radiative transport over convection. We can determine this point on the H-R diagram by considering the actual temperature gradient and comparing this with the radiative temperature gradient for a given $L(r)$ and κ.

The internal temperature varies as $T \propto GM/R$ as the star moves down the Hayashi track, so the actual temperature gradient is

$$\frac{\partial T}{\partial r} \propto \frac{GM}{R^2}. \tag{10.60}$$

The radiative temperature gradient is

$$\left.\frac{dT}{dr}\right|_{\rm rad} = -\frac{3\kappa \rho F}{16\pi a c T^3 r^2} \sim \frac{\kappa \left(M/R^3\right) L}{(\mu M R^3) R^2} \sim \frac{\kappa L}{\mu^3 M^2 R^2}. \tag{10.61}$$

Since $\kappa \propto \rho^n T^{-s} \sim M^{n-s} R^{s-3n}$, then

$$\left.\frac{dT}{dr}\right|_{\rm rad} \propto M^{n-s-2} L R^{s-3n-2}. \tag{10.62}$$

Now, when the radiative temperature gradient equals the actual temperature gradient, then radiative transfer will take over and convection will cease. This occurs when the ratio of these gradients equals one. This is equivalent to requiring

$$\frac{M^{s-n+3}}{L R^{s-3n}} = C, \tag{10.63}$$

where C is a constant. Since $T_{\rm eff}$ is essentially constant as the star moves along the Hayashi track, we assume $L \propto R^2$, so

$$C = \frac{M^{s-n+3}}{R^{s-3n+2}}, \tag{10.64}$$

and

$$\frac{d\ln L}{d\ln M} = 2\left(\frac{s-n+3}{s-3n+2}\right) \tag{10.65}$$

if $M^{s-n+3} \propto R^{s-3n+2}$.

From the equation of effective temperature, we have

$$\left.\frac{\partial \ln T_{\rm eff}}{\partial \ln M}\right|_L = \frac{n+3}{9n+3-2s}, \tag{10.66}$$

so the line of constant ratios has a slope of

$$\frac{\partial \ln L}{\partial \ln T_{\text{eff}}} = \frac{2(9n+3-2s)}{n+3} \frac{s-n+3}{s-3n+2}. \quad (10.67)$$

For H$^-$ opacity, this slope is 39. Accurately calculating these gradients for a specific case will set the location of this line. Once a collapsing star crosses this line, it will leave the Hayashi track as the core begins to become radiative. At this point, the luminosity evolution again follows from the conversion of gravitational energy, so

$$L \propto T_{\text{eff}}^{2.2}. \quad (10.68)$$

The track along this line is called the Henyey track. At this point, the core density increases and the core temperature increases until fusion begins and the collapse is halted. The star is now at the zero age main sequence (ZAMS). Very low mass stars may not achieve sufficient temperature and density in the core to ignite fusion. In this case electron degeneracy pressure may be sufficient to halt the collapse. These are called brown dwarfs.

Problem 10.4: Assume that the opacity is dominated by the H$^-$ ion. Show that

$$R \propto \left(\frac{M}{M_\odot}\right)^{13/17} R_\odot$$

when a star enters the Henyey track. If a $1\,M_\odot$ star has a radius of $5\,R_\odot$ when it enters the Henyey track, what is the radius of a $2\,M_\odot$ star at this point in its evolution?

The generic path of the birth of a star from collapse of the cloud to ZAMS is shown in Fig. 10.2. Stars with final masses below about $0.5\,M_\odot$ will achieve fusion in the core before leaving the Hayashi track. In general, the timescales are determined by the mass of the object, and so low-mass stars spend more time at every phase of the birthing process.

Problems

10.1. The Jeans length (R_J) is defined to be the minimum radius necessary to collapse a cloud of density ρ_0.

(a) Use the expression of the Jeans mass to obtain the following one for the Jeans length:

$$R_J = \sqrt{\frac{15\mathscr{R}T}{4\pi\mu G\rho_0}}.$$

Fig. 10.2 Rough sketch of the main processes outlined in this chapter

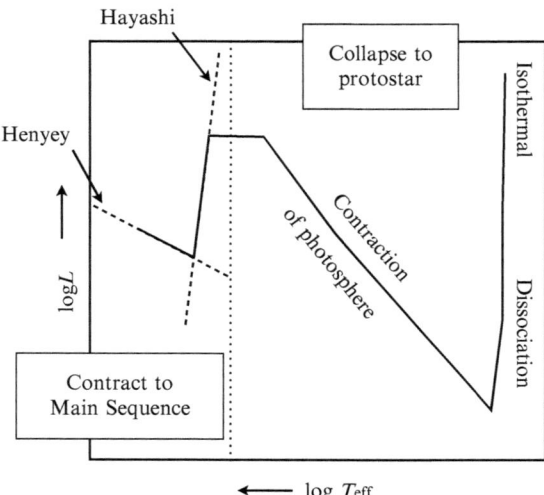

(b) For a typical diffuse hydrogen cloud, $T = 50\,\text{K}$, and $n = 5 \times 10^8\,\text{m}^{-3}$. If we assume that the cloud is entirely composed of H I, $\rho_0 = m_H n = 8.4 \times 10^{-19}\,\text{kg/m}^3$. Taking $\mu = 1$, determine R_J for this cloud.

10.2. Using the opacity for dust and taking the initial density to be $\rho \simeq 10^{-10}\,\text{kg/m}^3$, determine the radius of the star when the optical depth becomes $2/3$ at the center. Assume constant density.

10.3. The typical metal composition of a star with solar metallicity is described by $X = 0.68$, $Y = 0.3$, and $Z = 0.02$. Using these values, determine the radius and central temperature of the star after ionization of hydrogen and helium.

10.4. Assume that the opacity is dominated by the H^- ion. Show that

$$R \propto \left(\frac{M}{M_\odot}\right)^{13/17} R_\odot$$

when a star enters the Henyey track. If a $1\,M_\odot$ star has a radius of $5\,R_\odot$ when it enters the Henyey track, what is the radius of a $2\,M_\odot$ star at this point in its evolution?

Chapter 11
Main Sequence Structure

Numerical solutions to the equations of stellar structure can provide detailed description of the interiors of stars. Some general features of stellar structure can be discovered through these solutions. The mass distribution indicates that for high-mass stars with $M \geq 1.2\,\mathrm{M}_\odot$ over 95 % of the mass is contained within $\sim 60\,\%$ of the radius. For lower-mass stars, this extends out to about 80 % of the radius. Naturally, the energy production is significantly higher in high-mass stars and extends out to larger radii. Combined with the increased density in the cores of higher-mass stars, this increased energy production results in a convective core and a radiative envelope for stars with $M \geq 1.2\,\mathrm{M}_\odot$. The situation is reversed for low-mass stars with $M \leq 1.2\,\mathrm{M}_\odot$, which have a radiative core and a convective envelope. The convection in the envelope is due to the increased opacity caused by the lower temperatures of low-mass stars. In this chapter, we will discuss the zero age main sequence (ZAMS) structure and subsequent evolution of high- and low-mass stars.

11.1 High-Mass Stars

When hydrogen begins to fuse into helium in the core of a star, it is considered to be at ZAMS. For most high-mass stars, the high densities and pressures in the core cause the CNO cycle to dominate the hydrogen burning process and the core is convective. As the hydrogen burning progresses, the number of nuclei in the core decreases and the mass fractions, X and Y, change. From Sect. 5.1, we have

$$\frac{1}{\mu} = \frac{1}{\mu_I} + \frac{1}{\mu_e} = 2X + \frac{3}{4}Y + (1-X-Y)\left\langle \frac{1+\mathscr{Z}}{\mathscr{A}} \right\rangle, \qquad (11.1)$$

where we have used the fact that $X + Y + Z = 1$ to eliminate Z from the equation. Noting that

$$\left\langle \frac{1+\mathscr{Z}}{\mathscr{A}} \right\rangle \sim \frac{1}{2} \tag{11.2}$$

for lower-mass metals, we can write this as

$$\frac{1}{\mu} = \frac{1}{2} + \frac{3}{2}X + \frac{1}{4}Y. \tag{11.3}$$

Thus, as X decreases and Y increases, μ increases. During hydrogen burning, the gas at the core can be treated as an ideal gas, so

$$P = \left(\frac{\mathscr{R}}{\mu}\right)\rho T \tag{11.4}$$

and an increase in μ results in a decrease in the pressure. This decrease in pressure results in a contraction of the core and a subsequent rise in the density and temperature, restoring the pressure required to support the star. This is a consequence of the secular thermal stability discussed in Sect. 9.2. The increase of ρ and T also produces an increase in the nuclear burning rate, increasing the luminosity, L. Recalling the power-law approximations to the nuclear burning rates from Sect. 7.2, the CNO cycle depends on $T^{19.9}$, compared to the p-p chain which only depends on T^4. Consequently, there is a substantial increase in the luminosity for high-mass stars as a result of a small change in the core temperature.

Problem 11.1: Assume that the core contraction associated with the increase in μ is homologous, so that throughout the core, $r \to r + \delta r$. For an energy rate given by $q = q_0 \rho T^\beta$, show that the fractional change in energy rate is given by:

$$\frac{\delta q}{q} = -(3+\beta)\frac{\delta r}{r}.$$

Since the changes in the star are occurring on the nuclear timescale, we can assume that the star remains in both thermal and hydrodynamic equilibrium during these changes. Consequently, both the internal energy and the gravitational potential energy are separately conserved. Therefore, when the core shrinks, the envelope must expand to maintain the gravitational potential energy. Likewise, because the core temperature increases, the envelope must cool to maintain the internal energy. This produces the counterintuitive result that the contraction and heating of the core causes the surface of the star to expand and cool. Despite the increase in L_{nuc}, the effective temperature drops. Thus the star moves off to the right and above its ZAMS location during the main sequence evolution. This is a standard feature of stellar evolution that the radius, luminosity, and central temperature of a star increase during the main sequence phase.

11.1 High-Mass Stars

Fig. 11.1 The quantities associated with the calculation of the pressure in the isothermal core and the pressure from the weight of the envelope

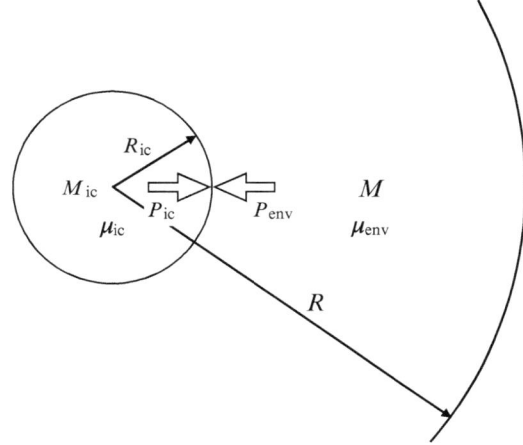

Eventually, the core hydrogen will be almost completely depleted and hydrogen burning stops in the core. For stars with convective cores, this occurs nearly simultaneously throughout the convective region in the core. When nuclear burning stops, the core quickly becomes isothermal as there are no new sources of heat in the core. In order to maintain hydrostatic pressure, the core must set up a pressure gradient to support the layers above it. From the ideal gas law, an isothermal core can only set up a pressure gradient by developing a density gradient, since $P \propto \rho T$. A generic feature of the cessation of core nuclear burning is a contraction of the core in order to produce this density gradient. The gravitational energy released through the collapse increases the luminosity, and the temperature at the outer edge of the core increases enough to ignite a hydrogen burning shell around the core. This causes the envelope to expand and the effective temperature to drop. The exact details of the ignition of shell burning hydrogen and the collapse of the helium core depend upon the ability of the isothermal core to support itself and the envelope above it. We will now look at the details of isothermal core equilibrium.

In hydrostatic equilibrium, the pressure supplied by the isothermal core as calculated from the core outward should equal the pressure caused by the weight of the envelope above it as calculated from the surface inward. By imposing this condition, we can obtain a constraint on the mass of the isothermal core, M_{ic}, as a fraction of the total mass M of the star. We define the mass fraction as

$$q = \frac{M_{\text{ic}}}{M}. \tag{11.5}$$

The basic configuration is shown in Fig. 11.1, and we will impose the condition $P_{\text{env}} = P_{\text{ic}}$.

We begin by calculating P_{env}, the pressure in the envelope from the equation of hydrostatic equilibrium:

$$P_{\text{env}} = \int_0^{P_{\text{env}}} dP = -\int_M^{M_{\text{ic}}} \frac{Gm\,dm}{4\pi r^4}. \tag{11.6}$$

The precise value of the integral depends upon the mass distribution in the envelope. Nonetheless, from dimensional considerations, we can approximate this integral by:

$$P_{\text{env}} \simeq \frac{-G}{8\pi \langle r^4 \rangle} \left(M_{\text{ic}}^2 - M^2\right), \tag{11.7}$$

where $\langle r^4 \rangle$ is the "weighted average" value of r^4 from the integral. Since we expect the isothermal core to be a small fraction of the total mass of the star we approximate $M^2 - M_{\text{ic}}^2 \sim M^2$, so

$$P_{\text{env}} \simeq \frac{GM^2}{8\pi \langle r^4 \rangle}. \tag{11.8}$$

In the absence of any compelling reason to do otherwise, we assume the average of r^4 is half the maximum value, so

$$P_{\text{env}} \simeq \frac{G}{4\pi} \frac{M^2}{R^4}. \tag{11.9}$$

At the boundary between the isothermal core and the envelope, the temperature of the gas in the envelope should equal the temperature of the isothermal core, so

$$P = \frac{k}{\mu m_{\text{H}}} \rho T \Longrightarrow T_{\text{ic}} = \frac{P_{\text{env}} \mu_{\text{env}} m_{\text{H}}}{\rho_{\text{env}} k}. \tag{11.10}$$

We approximate the density of the envelope at the boundary by the mean density, so

$$\rho_{\text{env}} \simeq \frac{3M}{4\pi R^3}, \tag{11.11}$$

and

$$R \simeq \frac{1}{3} \frac{GM}{T_{\text{ic}}} \frac{\mu_{\text{env}} m_{\text{H}}}{k}. \tag{11.12}$$

We can now turn this around and substitute this expression for R back into the equation for the hydrostatic pressure to obtain

$$P_{\text{env}} \simeq \frac{81}{4\pi} \frac{1}{G^3 M^2} \left(\frac{kT_{\text{ic}}}{\mu_{\text{env}} m_{\text{H}}}\right)^4. \tag{11.13}$$

This is the downward pressure of the envelope on the surface of the isothermal core.

The outward pressure from the isothermal core is calculated using the virial theorem. Recalling from Sect. 4.4, the virial theorem can be written as

$$d(PV) - \frac{P}{\rho}dm = -\frac{1}{3}\frac{Gmdm}{r}. \tag{11.14}$$

Integrating outward from the center to the surface of the isothermal core gives

$$3P_{ic}V_{ic} - 3\int_0^{M_{ic}} \frac{P}{\rho}dm = -\int_0^{M_{ic}} \frac{Gmdm}{r}, \tag{11.15}$$

where the subscript "ic" indicates the values at the surface of the isothermal core. The first term on the left-hand side of Eq. (11.15) is $4\pi R_{ic}^3 P_{ic}$. Assuming the core is an ideal gas, the second term is

$$3\int_0^{M_{ic}} \frac{P}{\rho}dm = 2U_{ic}, \tag{11.16}$$

where U_{ic} is the internal energy of the ideal gas, so

$$U_{ic} = \frac{3}{2}\left(\frac{M_{ic}}{\mu_{ic}m_H}\right)kT_{ic}. \tag{11.17}$$

On the right-hand side of Eq. (11.15), we have the gravitational potential of the isothermal core:

$$-\int_0^{M_{ic}} \frac{Gmdm}{r} = \Omega_{ic} \simeq -\frac{3}{5}\frac{GM_{ic}^2}{R_{ic}}, \tag{11.18}$$

where the factor of 3/5 arises from the assumption of constant density in the core.
Combining all of this together, we have for the pressure

$$P_{ic} = \frac{3}{4\pi R_{ic}^3}\left[\frac{M_{ic}kT_{ic}}{\mu_{ic}m_H} - \frac{GM_{ic}^2}{5R_{ic}}\right]. \tag{11.19}$$

The term in brackets is quadratic in M_{ic}, and so it has a maximum value at

$$M_{ic} = \frac{5}{2}\frac{kT_{ic}R_{ic}}{G\mu_{ic}m_H}. \tag{11.20}$$

This implies that the pressure of an isothermal core of mass M_{ic} and temperature T_{ic} has an upper bound, and so

$$P_{ic,max} = \frac{375}{64\pi}\frac{1}{G^3 M_{ic}^2}\left(\frac{kT_{ic}}{\mu_{ic}m_H}\right)^4. \tag{11.21}$$

Equating the maximum isothermal core pressure to the envelope pressure gives a critical value for the mass fraction of the isothermal core:

$$q_{\text{crit}} = \frac{M_{\text{ic}}}{M} = \sqrt{\frac{375}{1296}} \left(\frac{\mu_{\text{env}}}{\mu_{\text{ic}}}\right)^2 \simeq 0.54 \left(\frac{\mu_{\text{env}}}{\mu_{\text{ic}}}\right)^2. \tag{11.22}$$

This is known as the *Chandrasekhar–Schonberg limit*. A more rigorous calculation using the mass distribution for a $n = 1.5$ polytrope for a radiative envelope gives $q_{\text{crit}} = 0.37 \, (\mu_{\text{env}}/\mu_{\text{ic}})^2$.

Problem 11.2: A star starts out with a composition of $X = 0.7$, $Y = 0.26$, and $Z = 0.04$, the metals have a solar distribution. Assume all the hydrogen is burned in the core. What is the Chandrasekhar–Schonberg limit for this star?

If the isothermal core mass exceeds $0.37 \, (\mu_{\text{env}}/\mu_{\text{ic}})^2 M$, then the core cannot support the envelope and it begins to collapse. In some cases, the core can become partially degenerate prior to or during this collapse, and so the above analysis needs to be modified to handle the degeneracy.

Returning to the virial theorem in Eq. (11.15), we note that the internal energy is the only term that needs to be modified to include degeneracy. In order to approximate the smooth transition from classical to degenerate gas, we write the pressure as a combination of an ideal gas plus the contribution from degenerate electrons:

$$P = \frac{\rho}{\mu m_H} kT + K\rho^{5/3}, \tag{11.23}$$

where

$$K = \frac{1}{5} \frac{\hbar^2 \left(3\pi^2\right)^{2/3}}{m_e} \left[\frac{\mathscr{Z}}{\mathscr{A} m_H}\right]^{5/3}. \tag{11.24}$$

Recalling that

$$\frac{1}{\mu_e} = \sum_i \frac{X_i \mathscr{A}_i}{\mathscr{Z}_i} \tag{11.25}$$

and understanding that the composition of the core will change during the evolution, we can pull out the composition dependence by defining K_* such that

$$K = \frac{K_*}{\mu_e}. \tag{11.26}$$

Thus, we arrive at the approximation for the pressure of a partially degenerate gas:

$$P = \frac{\rho kT}{\mu m_H} + K_* \left(\frac{\rho}{\mu_e}\right)^{5/3}. \tag{11.27}$$

11.1 High-Mass Stars

Consequently, the internal energy integral in Eq. (11.15) becomes

$$3 \int_0^{M_{ic}} \frac{P}{\rho} dm = \frac{3M_{ic}kT_{ic}}{\mu_{ic}m_H} + 3K_* \int_0^{M_{ic}} \frac{\rho^{2/3}}{\mu_e^{5/3}} dm. \tag{11.28}$$

From dimensional considerations, the integral on the right-hand side can be written:

$$3K_* \int_0^{M_{ic}} \frac{\rho^{2/3}}{\mu_e^{5/3}} dm = \alpha \frac{3K_*}{\mu_e^{5/3}} \frac{M_{ic}^{5/3}}{R_{ic}^2} \tag{11.29}$$

where α is a proportionality constant of order unity that depends on the mass distribution in the isothermal core.

Therefore, the core pressure can be written:

$$P_{ic} = \frac{3}{4\pi R_{ic}^3} \left[\frac{M_{ic}kT_{ic}}{\mu_{ic}m_H} - \frac{1}{5} \frac{GM_{ic}^2}{R_{ic}} + \frac{\alpha K_*}{\mu_e} \frac{M_{ic}^{5/3}}{R_{ic}^2} \right]. \tag{11.30}$$

Partial degeneracy adds the third term in the brackets on the right-hand side. As a function of R_{ic}, the pressure now has two turning points at

$$R_{ic} = \frac{2}{15} \mu_{ic} \frac{GM_{ic}m_H}{kT_{ic}} \left[1 \pm \left(1 - \frac{375\alpha K_*}{4m_H G^2} \frac{kT_{ic}}{\mu_{ic}\mu_e M_{ic}^{4/3}} \right)^{1/2} \right]. \tag{11.31}$$

At the critical value of

$$M_{crit} = \left[\frac{375\alpha K_* kT_{ic}}{4G^2 \mu_{ic}\mu_e m_H} \right]^{3/4}, \tag{11.32}$$

the two turning points coincide at an inflection point. For higher masses, there are two turning points. For lower masses, there are none. For a core temperature of $T_{ic} \sim 2 \times 10^7$ K, the critical mass is $\sim 0.1 \, M_\odot$ for reasonable values of α. The pressure as a function of radius for values of M_{ic} around M_{crit} is shown in Fig. 11.2.

When hydrogen burning stops in the core and q is less than q_{crit}, then the isothermal core can support the envelope and the pressure and radius are given by a point on the $P(R)$ curve appropriate for the envelope pressure given by q. While hydrogen burning continues in a shell around the core, the mass ratio increases. An increasing mass ratio causes the $P(R)$ curve to move to that defined by higher core masses. If the core mass never exceeds M_{crit}, then the core smoothly transitions from ideal gas to degenerate gas. If the core mass eventually exceeds M_{crit}, then as the mass ratio increases, it will reach q_{crit}. At that point in the star's evolution, there is no stable solution and the core collapses toward the degenerate solution as the temperature increases.

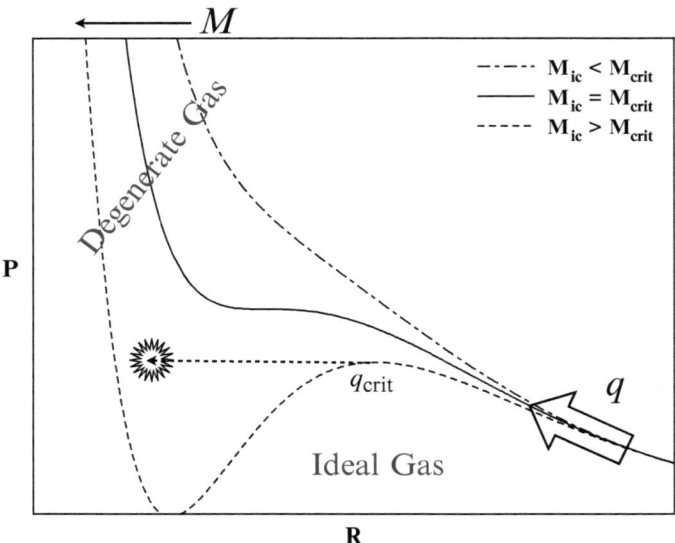

Fig. 11.2 The pressure as a function of radius for various isothermal core masses. The steep rise in pressure at small radii is due to the increasing degeneracy of the core gas. The star starts at a point on the curve at large radius. As q increases, the star moves up the pressure curve. For $M_{ic} < M_{crit}$, the star smoothly transitions to degeneracy. For $M_{ic} > M_{crit}$, q will eventually ready q_{crit}, and the core will rapidly collapse

Stars with $M \geq 3\,M_\odot$ have a core mass larger than M_{crit}. We will now follow the evolution of a high-mass star with shell hydrogen burning around an isothermal helium core. The shell burning increases the mass of the core until q exceeds q_{crit} and the core can no longer support the envelope. The ensuing collapse of the core occurs at a thermal timescale $t_{th} \sim GM^2/RL$. For typical core values of $M_{ic} \sim 0.1\,M_\odot$ and $R_{ic} \sim 0.3\,R_\odot$, the timescale is $t_{th} \sim 10^6$ years. In this case, energy conservation and the virial theorem require the gravitational potential energy, Ω, and the internal energy, U to be each conserved separately.

The self-energy of the core is substantially greater than the self-energy of the envelope, due to the smaller size and greater density of the core. Therefore, we can approximate the total potential energy of the star as the self-energy of the core plus the potential energy of the envelope due to the gravitational attraction of the core, and ignore the self-energy of the envelope. Thus, we have

$$|\Omega| \simeq \frac{GM_{ic}^2}{R_{ic}} + \frac{GM_{ic}M_{env}}{R} \sim \text{constant}, \qquad (11.33)$$

where R is the radius of the star. During the relatively rapid collapse of the core, shell burning does not have time to appreciably add to the mass of the core, so we can treat M_{ic} and M_{env} as constants. Conservation of Ω then gives

11.1 High-Mass Stars

$$\frac{d}{dt}|\Omega| = -\frac{GM_{ic}^2}{R_{ic}^2}\frac{dR_{ic}}{dt} - \frac{GM_{ic}M_{env}}{R^2}\frac{dR}{dt} = 0. \quad (11.34)$$

Therefore, we have

$$\frac{dR}{dt} = -\left(\frac{M_{ic}}{M_{env}}\right)\left(\frac{R}{R_{ic}}\right)^2\frac{dR_{ic}}{dt}, \quad (11.35)$$

and the radius of the star increases as the core contracts. During this phase, the matter below the hydrogen burning shell is collapsing while the matter above it is expanding. This stage where the bulk of the luminosity arises from the collapse of the core happens quickly (10^6 years) and so it is rare to observe stars at this point in their evolution off the main sequence over to the red giant phase. This region of the Hertzsprung–Russell diagram is called the *Hertzsprung gap*.

> **Problem 11.3** During the transition over the Hertzsprung gap, the radius of a $10\,M_\odot$ star increases from $R_i = 8\,R_\odot$ to $R_f = 250\,R_\odot$. Assume that the core has a mass of $M_{ic} = 0.1\,M_\odot$ and initial radius of $R_{0ic} = 0.3\,R_\odot$. Use Eq. (11.35) to show that the radius of the core after the collapse is $0.065\,R_\odot$.

As the envelope expands and cools, eventually the temperature drops to the point that H^- ions can form and begin to contribute to the opacity. The increased opacity results in the development of a convection zone in the outer atmosphere. The star continues to expand and cool due to the collapsing core, and so its effective temperature and luminosity follow a Hayashi track, but in the reverse direction from a collapsing protostar. Thus, the luminosity increases substantially while the effective temperature remains fairly constant. At this point the star is a red giant. The convection zone can reach all the way down to the hydrogen burning shell and transport material enriched in metals from nuclear burning in the shell up to the surface. This process is known as *dredge-up*.

The core remains nondegenerate during the collapse and as the temperature in the core increases, it eventually reaches the ignition temperature for the triple alpha process ($T \sim 10^8$ K). The contraction of the core then halts and the star settles into thermal and hydrodynamic equilibrium. The core expands slightly and the envelope contracts. The star now consists of a helium burning core surrounded by a hydrogen burning shell. As helium burning proceeds, the mean molecular weight increases in the core as three helium nuclei are combined to form one carbon nucleus. The change is not as dramatic as for hydrogen burning, but the core does contract slightly and the luminosity increases.

During the helium burning stage, the convective region in the envelope shrinks toward the surface, but the high-temperature dependence of the triple alpha process causes a convective core to again develop. Stars burning helium in their cores occupy a region in the Hertzsprung–Russell diagram called the *horizontal branch*. Stars proceed through the horizontal branch on a nuclear timescale, so they are frequently seen. Eventually a carbon core develops with a helium burning shell. Once again,

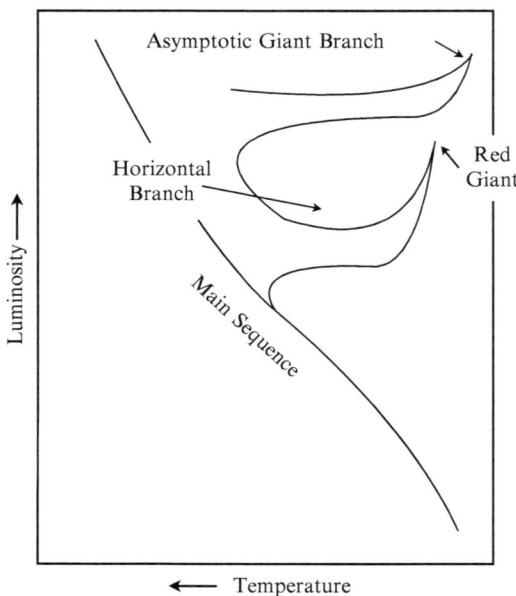

Fig. 11.3 Sketch of a typical evolutionary path of a star through to the formation of a carbon–oxygen core

we have a contracting core surrounded by a burning shell and so the envelope expands and the luminosity drops. The evolution is now similar to the evolution toward the red giant phase, except there are two shells of burning—an inner one of helium and an outer one of hydrogen. A convective envelope develops and there is a second dredge-up. The star follows the Hayashi track again along what is called the *asymptotic giant branch (AGB)*. A typical track of a massive star in the Hertzsprung–Russell diagram is shown in Fig. 11.3. Further evolution depends strongly on the mass of the star and can end in either a white dwarf or a supernovae. Before we cover the details of late-stage evolution, we will look at low-mass stars.

11.2 Low-Mass Evolution

On the main sequence, low-mass stars have radiative cores and convective envelopes. This has consequences as the core hydrogen is burned into helium. In contrast to high-mass stars where convection ensures that fresh hydrogen is mixed throughout the core, low-mass stars develop a hydrogen gradient and so the transition from core burning to shell burning is gradual. Furthermore, the temperature sensitivity of the p-p chain is lower, so the increased temperature at the core due to changes in the mean molecular weight does not result in significant changes in the radius of the star. Thus, during the core hydrogen burning of the star, there is only a slight increase in T_{eff} and L. Once the hydrogen is exhausted

11.2 Low-Mass Evolution

in the center of the core, it begins to contract and shell burning slowly moves out from the core. This causes the envelope to expand, but with very little change in luminosity, so the effective temperature drops. As the mass of the isothermal helium core grows, the core becomes nearly degenerate while supporting the hydrogen shell. Thus, there is no rapid collapse of the core as there is in higher-mass stars. Low-mass stars simply move on a nuclear timescale from main sequence to red giant with a degenerate helium core and a hydrogen burning shell. Consequently, there is no Hertzsprung gap for low-mass stars.

As core contraction proceeds, the temperature and densities will eventually reach the values where the triple alpha process can ignite. This occurs when the core mass is $\sim 0.45 M_\odot$ and $T_c \sim 10^8$ K. At this point, however, the core is fully degenerate, and the equation of state

$$P \propto \rho^{5/3} \tag{11.36}$$

is independent of temperature. Thus, the onset of nuclear burning is thermally unstable and does not cause the core to expand and cool. Instead, the temperature increases and the nuclear burning increases with it until the degeneracy is lifted. The unstable runaway of helium burning is known as the *helium flash*.

The helium flash describes the final few days of helium burning in a degenerate core, prior to the lifting of the degeneracy. During this time the nuclear luminosity of the core rises as high as $L_{\text{nuc}} > 10^9 L_\odot$. This burst of luminosity does not manifest itself as an explosive event. Instead, the helium core slowly expands while convection, conduction, and radiation smoothly deliver the energy to the surface. Detailed two- and three-dimensional hydrodynamic simulations are needed to model the helium flash. The energy production rate for a simple one-dimensional model is shown in Fig. 11.4.

Once the degeneracy is lifted, then further increases in T result in an expansion of the core and the burning becomes stable. The expansion of the core results in a shrinking of the envelope and there is a reduction in the luminosity of the star as it stably burns helium into carbon and oxygen. If a star is sufficiently low mass, then the expansion to the red giant phase is accompanied by an increased stellar wind and mass loss from the envelope. This is caused by the increased opacity of the stellar envelope and the increased luminosity from the contracting helium core. Since the expanded envelope is also less gravitationally bound to the core, low-mass stars may lose their entire envelope before the core reaches temperatures high enough to ignite helium burning. The degenerate helium core simply continues to contract and cool, resulting in a helium white dwarf. Stars with masses low enough to result in helium white dwarfs have very long lifetimes and have not had the opportunity to reach this stage in their lives within the age of the universe. Any helium white dwarfs that are observed in the universe today must have followed a different evolutionary path than the isolated stellar evolution described here. We will see in Chap. 13 how helium white dwarfs arise from the evolution of interacting binary stars.

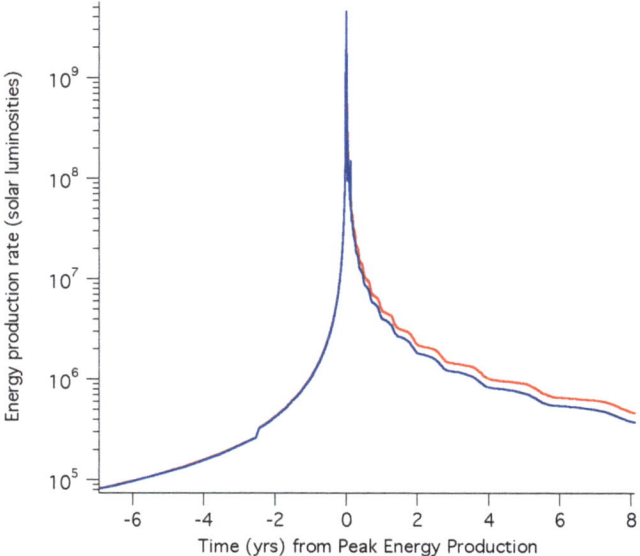

Fig. 11.4 The time history of the energy production rate from helium burning as calculated from a one-dimensional model of Dearborn, Lattanzio, and Eggleton. The rate of change of the thermal energy of the core is shown in blue. Figure from Dearborn, Lattanzio and Eggleton, *Astrophysical Journal*, 639, 405 (2006). Reproduced by permission of the AAS

11.3 Late-Stage Evolution

At this point, we have evolved both high- and low-mass stars up to the onset of helium burning. So far, the main difference between high- and low-mass evolution is that high-mass stars tend to smoothly transition to helium burning while low-mass stars undergo a dramatic flash event that injects a pulse of energy into the envelope. This trend continues through the subsequent evolution. This time we will start with low-mass stars and conclude with high-mass stars.

For low-mass stars, the burning of helium into carbon and oxygen proceeds until a carbon–oxygen core develops. With the loss of energy, the carbon–oxygen core becomes isothermal and begins to collapse. Repeating the process at the creation of the helium core, the collapse of the carbon–oxygen core results in an expansion and cooling of the envelope of the star. The star grows even larger and is found in the Hertzsprung–Russell diagram on what is called the *AGB*. These are known as AGB stars. The expansion of the envelope at first includes the hydrogen burning shell around the helium core, and so hydrogen burning is extinguished as the shell cools. A second convection zone develops that reaches down to the upper regions of the now quiescent hydrogen burning shell. This *second dredge-up* transports helium and nitrogen to the surface of the star.

11.3 Late-Stage Evolution

The structure of an AGB star consists of a degenerate carbon–oxygen core, surrounded by a shell of helium, that is surrounded by an envelope of hydrogen. Because of the different energy generation rates and the response of the star to the changing core, nuclear burning in the shells undergoes a series of pulses, with burning oscillating between the hydrogen and helium shells. Hydrogen burning in the outer shell leads to the deposition of helium onto the quiescent helium shell, which contracts and heats up due to the addition of mass. As mass is added to the helium shell, it eventually reaches the conditions necessary for the helium to ignite. Burning in the helium shell is subject to the thin-shell instability discussed in Chap. 9, and so ignition is accompanied by explosive burning called a *shell flash*. The rapid release of energy during the shell flash causes the outer hydrogen burning shell to expand and cool—turning off the hydrogen burning shell. After the shell flash, the helium burning continues through the helium shell, depositing the carbon–oxygen ashes onto the core until the helium shell is depleted. At this point, the hydrogen shell reignites and begins to refill the helium shell.

The high luminosity of the shell flash can also push off the outer layers of the star. We can see this by noting that the maximum convective luminosity is set by the speed of sound. If a convective cell tries to transport matter faster than the speed of sound (c_s), then a shock wave will develop. Therefore, the maximum convective luminosity is

$$L_{\text{max, c}} \simeq \left(4\pi r^2\right) U c_s, \tag{11.37}$$

where $U \propto \rho kT/m_H$ and $c_s \simeq (kT/\mu m_H)^{1/2}$, so

$$L_{\text{max, c}} \simeq 4\pi r^2 \rho \mu \left(\frac{kT}{\mu m_H}\right)^{3/2}. \tag{11.38}$$

If the luminosity at the base of the envelope exceeds this value, then the radiative transport takes over at the base of the envelope. The pressure gradient set up by this is

$$\frac{dP}{dr} = \frac{dP_{\text{rad}}}{dr} + \frac{dP_{\text{gas}}}{dr} = \frac{4a}{3}T^3\frac{dT}{dr} + \frac{kT}{\mu m_H}\frac{d\rho}{dr} + \frac{k\rho}{\mu m_H}\frac{dT}{dr}. \tag{11.39}$$

We know the radiative luminosity in terms of the temperature gradient as

$$L = -\frac{16\pi ac}{3\kappa\rho}r^2 T^3 \frac{dT}{dr}, \tag{11.40}$$

so

$$\frac{P_{\text{gas}}}{\rho}\frac{d\rho}{dr} = \frac{1}{r^2}\left[\left(\frac{\kappa \rho L}{4\pi c}\right)\left(\frac{3P_{\text{gas}}}{4P_{\text{rad}}}+1\right) - mG\rho - \rho r^2 \ddot{r}\right]. \tag{11.41}$$

If hydrostatic equilibrium is maintained, then $\ddot{r} = 0$ and $d\rho/dr < 0$, so at the base of the envelope (at radius r_1), we must have

$$\frac{\kappa\rho L(r_1)}{4\pi G c m(r_1)}\left(\frac{3P_{\text{gas}}}{4P_{\text{rad}}}+1\right) \leq 1 \qquad (11.42)$$

or

$$L(r_1) \leq \frac{4\pi G c m(r_1)}{\kappa\rho}\left(\frac{3P_{\text{gas}}}{4P_{\text{rad}}}+1\right)^{-1}. \qquad (11.43)$$

If the shell flash luminosity exceeds this value then $\ddot{r} > 0$ at the base of the envelope and the envelope accelerates away. If the flash is strong then some of the envelope can be accelerated up to the escape velocity and be ejected. On the other hand, if the shell flash is weak, then the envelope can expand enough that opacity drops and the outer envelope cools sufficiently that there is a pressure drop and the envelope recollapses. The shock front that develops when the envelope falls back on the core can also eject outer layers of the envelope. Numerical simulations indicate that AGB envelopes can be ejected in this fashion on timescales of a few thousand years.

Another mechanism by which stars can lose mass is through a stellar wind. The details of stellar winds are not well understood, but the gross features can be understood from an interplay of the expansion of the atmosphere and the increased heating of the core. The expanding atmosphere cools to the point that molecules and dust particles begin to coalesce in the outer layers of the envelope. These particles are accelerated by the radiation pressure from the increasingly hot core. If they achieve escape velocity, they will entrain gas about them and carry away some of the atmosphere as well. This reduces the mass of the envelope and consequently the escape velocity is also lowered, making it easier for more mass to escape. Although the details are sketchy, the end result is clear. The loss of the envelope leaves a naked carbon–oxygen core with a thin atmosphere of either helium or hydrogen, depending on the last stage of the shell burning cycle. The core is surrounded by a shell of the ejected gas that fluoresces due to the strong ultraviolet emission from the hot core. These are known as planetary nebulae. The naked core eventually cools off to become a white dwarf. Carbon–oxygen white dwarfs are the end products of stars with initial masses of $M \leq 10 M_\odot$.

Higher-mass stars smoothly transition from helium burning to carbon burning, maintaining shell burning in both the helium shell and the hydrogen shell. This is due to the fact that the core mass is greater than the critical mass described in Sect. 11.1 and so collapse proceeds to ignition before the core becomes degenerate. This same process is repeated when the carbon is depleted in the core, leading to neon and oxygen burning. Successively heavier nuclei are fused in the core of the star until a core of nickel and iron develops. Despite the presence of nickel and other heavy elements, this is referred to as an *iron core*. The typical structure of a massive star at this point in its evolution is shown in Fig. 11.5.

Recalling from Chap. 7, the fusion of higher-mass nuclei yields less energy per mass and so the burning rates must increase in order to supply the needed energy to support the star. The last phase of silicon burning typically lasts only a few days. Further fusion in the iron core is endothermic and cannot supply any extra energy.

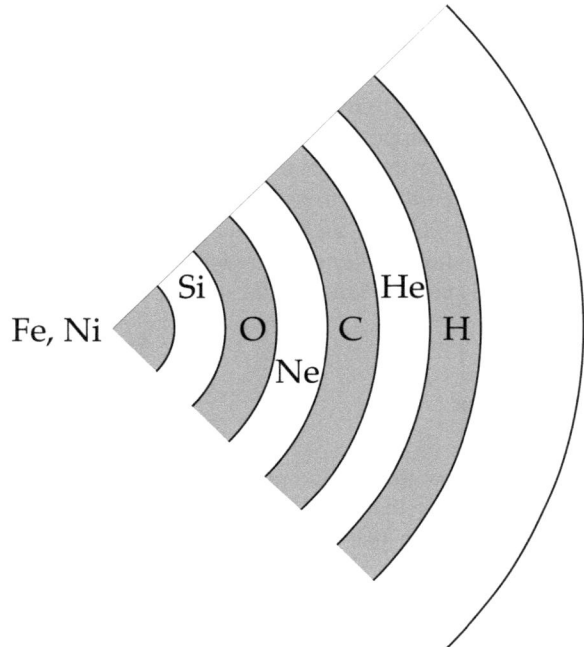

Fig. 11.5 Schematic of the nuclear burning shells in a high-mass star at the development of the iron core. Burning is proceeding in each shell, fusing elements in the shell to feed the shell below it. The relative sizes of the shells are not to scale and the size of the envelope would be substantially larger than shown

Therefore, once the iron core reaches the critical mass ratio and begins to collapse, no new source of energy is available to halt the collapse. The increased density and temperature lead to photodissociation of the heavy nuclei, returning the core to a collection of protons, neutrons, and electrons. Degenerate electron pressure is insufficient to halt the collapse once the Chandrasekhar mass is reached. At sufficiently high densities, the process of inverse beta decay occurs as the electrons and protons combine to form neutrons. All of these processes continue to remove energy from the core during further collapse. Additionally, the inverse beta decay removes electrons from the core, eliminating any degenerate electron pressure and lowering the total number of particles in the gas, which further drops the pressure.

The net result is a catastrophic loss of pressure and support for the core and it collapses. The collapse is homologous, so the inward velocity of each layer in the core is proportional to its radius. This means that there exists a radius where the collapse velocity exceeds the speed of sound in the gas and the core decouples from the envelope. The upper envelope falls in with free-fall velocity.

The collapsing core continues to shrink until the central density surpasses typically nuclear densities and the neutrons become degenerate. The degenerate pressure halts the collapse. The core rebounds slightly and ends at a radius of about 10–20 km. When the falling envelope strikes the rebounding core, an outward traveling shock wave develops. However, the shock is eventually overcome by the envelope and stalls, becoming an accretion shock, similar to the one encountered in the formation of a protostar. A major difference between star formation and this case is a strong outflow of neutrinos coming from the core. The shock front is sufficiently opaque to neutrinos so that the momentum added from absorption of these neutrinos can push the envelope off of the core and unbind it in a supernova explosion.

The remaining core will emerge as a neutron star if the mass of the core is low enough to be supported by degenerate neutron pressure. If the core mass exceeds a certain value that depends upon the equation of state for neutrons at nuclear densities, then the neutrons become relativistic and the core collapses to a black hole. For sufficiently massive stars, particle pair production during the evolution can cause an instability that disrupts the star before any degenerate core develops, and there is nothing left after the supernova.

Problems

11.1. Assume that the core contraction associated with the increase in μ is homologous, so that throughout the core, $r \to r + \delta r$. For an energy rate given by $q = q_0 \rho T^\beta$, show that the fractional change in energy rate is given by

$$\frac{\delta q}{q} = -(3+\beta)\frac{\delta r}{r}.$$

11.2. A star starts out with a composition of $X = 0.7$, $Y = 0.26$, and $Z = 0.04$, the metals have a solar distribution. Assume all the hydrogen is burned in the core. What is the Chandrasekhar–Schonberg limit for this star?

11.3. During the transition over the Hertzsprung gap, the radius of a $10 M_\odot$ star increases from $R_i = 8 R_\odot$ to $R_f = 250 R_\odot$. Assume that the core has a mass of $M_{ic} = 0.1 M_\odot$ and initial radius of $R_{0ic} = 0.3 R_\odot$. Use Eq. (11.35) to show that the radius of the core after the collapse is $0.065 R_\odot$.

Chapter 12
Compact Remnants

White dwarfs, neutron stars, and black holes are the compact remnants that are the endpoints of stellar evolution. Although nuclear processes have stopped in these objects, they each continue to evolve in ways that are unique to their structure. As white dwarfs cool and become more degenerate, their baryon structure changes, altering their cooling process. Neutron stars are frequently born with high magnetic fields that slowly decay and spin down the neutron star. Black holes, however, will not change much for eons. In this chapter, we will look at the properties and continued evolution of these exotic objects.

12.1 White Dwarfs

A white dwarf is supported by degenerate electron pressure, and so its structure can be approximated quite well by a polytrope of index $n = 1.5$ if the electrons are nonrelativistic. In this case, the mass–radius relation scales as $R \propto M^{-1/3}$. The typical radius can be found by noting that $R = \alpha R_n$, where α is found from Eq. (8.21):

$$\alpha = \left[\frac{(n+1)K}{4\pi G}\rho_c^{(1-n)/n}\right]^{1/2}, \tag{12.1}$$

and K is found from Eq. (5.44):

$$K = \frac{1}{5}\frac{\hbar^2(3\pi^2)^{2/3}}{m_e}\left[\frac{\mathscr{Z}}{\mathscr{A}}\frac{1}{m_H}\right]^{5/3}. \tag{12.2}$$

We can relate the central density ρ_c to the average density (and thus to the mass M and radius R of the white dwarf) by

$$\rho_c = D_n\bar{\rho} = D_n\frac{3M}{4\pi R^3}. \tag{12.3}$$

M. Benacquista, *An Introduction to the Evolution of Single and Binary Stars*, Undergraduate Lecture Notes in Physics, DOI 10.1007/978-1-4419-9991-7_12, © Springer Science+Business Media New York 2013

Fig. 12.1 Approximate structure of a white dwarf showing the atmosphere and the core. The thickness of the atmosphere is greatly exaggerated

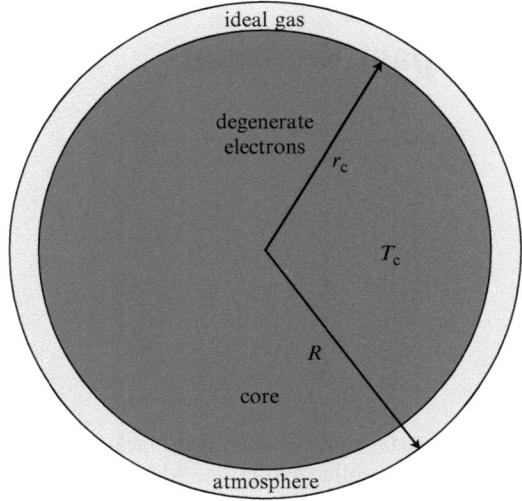

Combining these together yields the radius as a function of the mass:

$$R = \frac{(3\pi^2)^{1/3}}{16} \frac{\hbar^2}{Gm_e m_H^{5/3}} \frac{R_n^2}{D_n^{1/3}} \frac{1}{M^{1/3}}, \quad (12.4)$$

where R_n and D_n come from the numerical solution for the $n = 1.5$ polytrope. The radius for a typical white dwarf of mass $M = 0.7 M_\odot$ is found to be about $R = 10,000$ km, or nearly twice the size of the earth.

Near the surface, the density drops to zero and so there must be a nondegenerate atmosphere of the white dwarf. The thickness of this atmosphere is on the order of 10^{-3} of the white dwarf radius, and so it is a very thin surface layer. This thin layer of non-degenerate matter provides an insulating layer that regulates the rate of heat loss from the white dwarf.

The rate of heat loss through the atmosphere can be approximated by assuming that the white dwarf consists of a degenerate core of radius r_c surrounded by a layer of an ideal gas that extends out to the surface at radius R. Although the transition from degenerate to ideal gas should be smooth and gradual, we approximate the transition as a discrete step at r_c. The approximated structure of a cooling white dwarf is shown in Fig. 12.1.

Due to the high conductivity of the degenerate electrons, the core can be considered to be isothermal with temperature T_c. The temperature at the base of the atmosphere must also be T_c so that the temperature is continuous. As there are no additional sources of energy in the atmosphere, the heat flux is constant throughout the atmosphere and is equal to the luminosity L of the white dwarf. Therefore, we can determine the temperature and luminosity evolution of the white dwarf by considering the luminosity as a function of the core temperature T_c and relate this to

12.1 White Dwarfs

the heat capacity of degenerate core. We can use the hydrostatic equation [Eq. (8.1)] and the heat equation [Eq. (8.3)] combined with a Kramers opacity law to obtain an expression relating the pressure and temperature throughout the atmosphere.

We assume that the mass of the atmosphere is negligible and so $m = M$ throughout the atmosphere, and we have

$$\frac{dP}{dm} = -\frac{GM}{4\pi r^4}. \tag{12.5}$$

The heat equation reads

$$\frac{dT}{dm} = -\frac{3}{4ac}\frac{\kappa}{T^3}\frac{L}{(4\pi r^2)^2}, \tag{12.6}$$

with a Kramers opacity given by

$$\kappa = \kappa_0 \rho T^{-7/2}. \tag{12.7}$$

Combining these using

$$dP = \frac{dP}{dm}\frac{dm}{dT}dT \tag{12.8}$$

gives

$$dP = \frac{16\pi GacMT^{13/2}}{3\kappa_0 \rho L}dT. \tag{12.9}$$

We can eliminate the density from this equation by using the ideal gas law:

$$\rho = \frac{P\mu}{\mathcal{R}T}, \tag{12.10}$$

where μ is the mean molecular weight of both electrons and ions in the atmosphere. Thus, we have

$$PdP = \frac{16\pi Gac\mathcal{R}}{3\kappa_0 \mu}\left(\frac{M}{L}\right)T^{15/2}dT. \tag{12.11}$$

Integrating both sides of this equation from the surface of the white dwarf down to any depth between R and r_c gives an equation describing the pressure as a function of the temperature, mass, and luminosity of the white dwarf:

$$P(T) = \left[\frac{64\pi Gac\mathcal{R}}{51\kappa_0 \mu}\frac{M}{L}\right]^{1/2}T^{17/4}. \tag{12.12}$$

At the boundary where $r = r_c$, the temperature must be continuous, so $T = T_c$. Since the atmosphere is in hydrostatic equilibrium, the pressure in the atmosphere must be equal to the degenerate pressure in the core. The ions are not degenerate and so the ion pressures are identical on either side of the boundary. Therefore, equating the pressures is equivalent to setting the degenerate electron pressure equal to the ideal gas pressure due to the electrons:

$$K\rho^{5/3} = \frac{\mathscr{R}}{\mu_e}\rho T_c. \tag{12.13}$$

This provides a relation between the density of the atmosphere at the boundary and the temperature of the core which can be used to find an additional relationship between P and T_c:

$$P = \frac{1}{\mu^2\mu_e^3}\frac{\mathscr{R}}{K^3}T_c^5. \tag{12.14}$$

Combining this with Eq. (12.12) gives the luminosity of the white dwarf as a function of its mass and core temperature:

$$\frac{L}{M} = \left[\frac{64\pi G a c \mu \mu_e^3 K^3}{51\kappa_0 \mathscr{R}^4}\right]T_c^{7/2}. \tag{12.15}$$

For typical white dwarfs, the atmosphere consists of a thin layer of hydrogen above a layer of helium with some carbon and oxygen mixed in. The dominant source of opacity is then bound–free scattering, so

$$\kappa_0 = 4\times 10^{21}Z(1+X)\frac{m^5 K^{7/2}}{kg^2}. \tag{12.16}$$

Although there is hydrogen in the atmosphere, the mass fraction is almost negligible and so we adopt typical mass fraction values of $X = 0$, $Y = 0.95$, and $Z = 0.05$. In this case, $\kappa_0 = 2\times 10^{20}$. Assuming that the contribution to Z comes from equal parts carbon and oxygen, the mean molecular weight for ions is

$$\frac{1}{\mu_I} = X + \frac{Y}{4} + \frac{Z}{\langle\mathscr{A}\rangle} = \frac{0.95}{4} + \frac{0.05}{14} = 0.24, \tag{12.17}$$

where we have set $\langle\mathscr{A}\rangle = (12+16)/2$. This leads to $\mu_I = 4.15$. The mean molecular weight for the electrons is $\mu_e = 2$ because we have assumed $X = 0$ and there are no high $\mathscr{A}$ metals in the atmosphere. Therefore, the mean molecular weight is

$$\frac{1}{\mu} = 0.24 + 0.5 = 0.74 \Longrightarrow \mu = 1.35. \tag{12.18}$$

For $\mu_e = 2$, the degenerate equation of state constant is $K = 3.16\times 10^6$. Combining these typical white dwarf values with the other physical constants in Eq. (12.15), we find

$$\frac{L}{M} \simeq 1.1\left(\frac{T_c}{10^8\,\text{K}}\right)^{7/2}\frac{L_\odot}{M_\odot}. \tag{12.19}$$

12.1 White Dwarfs

The source of this luminosity is the gradual cooling of the core. Since the degenerate core is nearly incompressible, the cooling process can be modeled as a constant volume process, and so

$$L = -MC_V \frac{dT_c}{dt}. \tag{12.20}$$

The specific heat at constant volume can be split into two parts—one part due to the ideal gas of ions and the other part due to the fraction of degenerate electrons that can absorb or release thermal energy. The first part is quite simple:

$$C_V^{\text{ion}} = \frac{3}{2} \frac{k}{\mu m_H} = \frac{3}{2} \frac{\mathscr{R}}{\mu}. \tag{12.21}$$

The contribution of electrons to the specific heat is somewhat more complicated, but it can be approximated by noting that only those electrons with energy within kT of the Fermi energy can absorb and release thermal energy. Thus, we simply scale the specific heat to reflect this fraction:

$$C_V^e = \frac{3}{2} \frac{k}{\mu_e m_H} \frac{kT_c}{E_F}. \tag{12.22}$$

From this, we can see that although electron cooling can be important at the birth of a white dwarf, the core is degenerate and it is cooling; therefore the contribution of the electrons to the specific heat quickly becomes negligible as $kT_c \ll E_F$ and grows progressively less important. Therefore, we will approximate the specific heat as C_V^{ion}. Thus, the luminosity can be related to the cooling of the core by

$$\frac{L}{M} = -\frac{3}{2} \frac{\mathscr{R}}{\mu} \frac{dT_c}{dt}. \tag{12.23}$$

The temperature of the isothermal core is not an observable quantity for a white dwarf. The important quantity is the luminosity. Using Eq. (12.15), we can relate dT/dt to dL/dt, which we can then substitute into Eq. (12.23) and obtain a differential equation governing the time dependence of the white dwarf luminosity:

$$\frac{dL}{dt} \propto L^{12/7}. \tag{12.24}$$

The solution to this equation is

$$L(t) = L_0 \left[1 + \left(\frac{L_0}{M}\right)^{5/7} \frac{t}{\tau} \right]^{-7/5}, \tag{12.25}$$

where τ is the characteristic cooling time given by

$$\tau = \frac{3\mathscr{R}}{5\mu} \left[\frac{51\kappa_0 \mathscr{R}^4}{64\pi Gac\mu\mu_e^3 K^3} \right]^{2/7}. \tag{12.26}$$

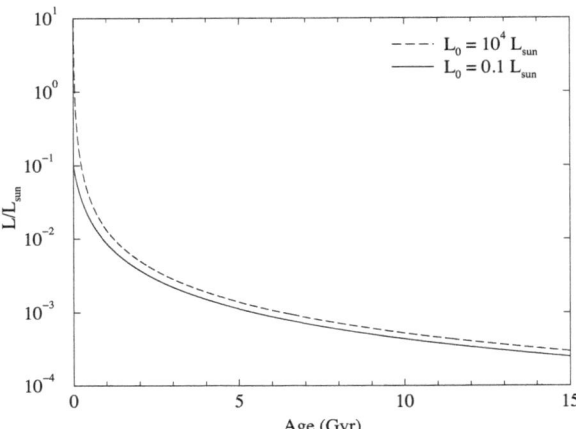

Fig. 12.2 Simple white dwarf luminosity evolution for a $0.7\,M_\odot$ white dwarf with a carbon–oxygen core and an atmosphere composition of $Y = 0.95$ and $Z = 0.05$. The solid line gives the evolution for an initial luminosity of $0.1\,L_\odot$, while the dashed line gives the evolution for an initial luminosity of $10^4\,L_\odot$. This model ignores any contribution to the luminosity arising from heat released during crystallization

For the typical white dwarf conditions described above, $\tau \sim 60\,\mathrm{Myr}$. This is a substantial underestimate at later times because the cooling time grows as the temperature decreases, but it shows that the initial cooling is quite rapid. The luminosity evolution for a $0.7\,M_\odot$ white dwarf is shown in Fig. 12.2. As can be seen there is a rapid drop in the luminosity in the first few hundred million years, after which the white dwarf luminosity stays around $0.001\,L_\odot$ for billions of years. Because the initial drop in luminosity is so rapid, the choice initial luminosity is almost unimportant in describing the long-term evolution of white dwarf luminosity.

> **Problem 12.1:** Calculate the effective temperature and the core temperature of a $0.7\,M_\odot$ white dwarf if its luminosity is $10^{-3}\,L_\odot$.

As the core of the white dwarf cools, the thermal energy of the ions will eventually become comparable to the energy of the electrostatic repulsion between the ions. At this point the ideal gas law will no longer be a valid description of the core. When this occurs, the ions will settle down into a crystal lattice with residual thermal energy of the ions appearing as vibrations about their lattice positions. This process is known as *crystallization* and is similar to a phase transition to a solid and so there is a release of a latent heat. The additional source of heat slows the cooling process at low luminosities. Once the core has become crystallized, coherent vibrations known as phonons then accelerate the cooling process at very low temperatures. The ultimate fate of a carbon–oxygen white dwarf is a large, dark carbon crystal.

One of the most notable features of white dwarfs is their faintness. From the cooling curves, most white dwarfs will have luminosities on the order of $10^{-3}\,L_\odot$. This corresponds to an absolute magnitude of about $M = 12$. At distances of 5 kpc, typical of nearby globular clusters, the apparent magnitude is 25.7. White dwarfs in the Magellanic clouds would be at the limiting magnitude for the Hubble space telescope.

12.2 Neutron Stars

Table 12.1 Spectral classification of white dwarfs

Spectral type	Spectral features
DA	Hydrogen balmer lines; no He or metals
DB	HeI lines; no H or metals
DC	Blackbody only; no lines
DO	HeII lines; HeI and/or H lines may be present
DZ	Metal lines only; no H or HeI lines
DQ	Carbon lines present
DX	Otherwise unclassifiable

Since they shine entirely by cooling, their position in the H-R diagram is found from $L = 4\pi\sigma R^2 T_{\text{eff}}^4$. Using $\log L$ and $\log T$ as variables, we have

$$\log L = \log 4\pi\sigma R^2 + 4\log T. \tag{12.27}$$

Because a white dwarf of a given mass has a specific radius, white dwarfs lie along lines of slope 4 in the H-R diagram, with intercepts determined by the mass. The spectrum of a white dwarf is dominated by the blackbody spectrum with spectral lines indicating the composition of the atmosphere. White dwarfs are classified by the dominant lines in their absorption spectra. The spectral classes are given in Table 12.1

12.2 Neutron Stars

When the core of a massive star collapses, the subsequent explosion is called a type II supernova. One of the possible remnants of a supernova is a neutron star—a compact object of around $1.4\,M_\odot$ that is supported by degenerate neutron pressure. We can estimate the size of a typical neutron star by replacing m_e and m_H in Eq. (12.2) with m_n and setting $\langle \mathscr{Z}/\mathscr{A} \rangle = 1$ to obtain

$$R_{\text{ns}} = \left(\frac{3\pi^2}{128}\right)^{1/3} \frac{\hbar^2}{G} \frac{R_n^2}{m_n^{8/3}} \frac{1}{(D_n M)^{1/3}}. \tag{12.28}$$

For $M = 1.4\,M_\odot$, this gives $R_{\text{ns}} = 13.5\,\text{km}$. This is a remarkably dense object with over a solar mass concentrated in a volume the size of a small city. The escape velocity for such an object is

$$v_{\text{esc}} = \sqrt{\frac{2GM}{R}} = 0.55\,c. \tag{12.29}$$

This clearly indicates that relativistic effects will be important when discussing neutron stars. The gravitational potential energy of an object at the surface is about

15 % of its rest mass. Nonetheless, the assumption of a nonrelativistic degenerate equation of state is still valid for a typical $1.4\,M_\odot$ neutron star, since $E_F \sim 0.03 m_n c^2$.

The average density of a $1.4\,M_\odot$ neutron star is $2.7 \times 10^{17}\,\text{kg/m}^3$, which is slightly higher than the density of the typical atomic nucleus. Using a $n = 1.5$ polytrope model, the central density is then $\rho_c = D_n \bar\rho = 1.6 \times 10^{18}\,\text{kg/m}^3$. There are few observations of matter under these conditions and the full equation of state is unknown for matter at densities above the nuclear density. Thus, our understanding of the conditions in the centers of neutron stars is still uncertain. Nonetheless, we expect the core of a neutron star to be a mixture of neutrons, pions, other baryons and mesons, and possibly a quark-gluon plasma. As we move outward from the core, we encounter the bulk of the neutron star interior, which consists of degenerate neutrons that supply the necessary pressure to support the star. In addition to the degenerate neutrons, there are a few protons and electrons whose relative abundances are determined by the equilibrium state of the reaction $n \rightleftharpoons p + e^-$, taking into account the degeneracy of the electrons and protons. Despite the relatively high temperature of neutron stars ($\sim 10^6\,\text{K}$), the neutrons and protons can be considered cold compared to the Fermi temperature. Thus, they form Cooper pairs in a process similar to the behavior of electrons in a typical low-temperature superconductor. This results in a fluid of neutrons and protons which is a superconducting superfluid. In a superconductor, electric current flows without any resistance. In a superfluid, the fluid itself flows without any viscosity. One effect of superfluidity and superconductivity is to freeze in any magnetic fields that are present in the interior of a neutron star at its creation and to lock in much of the internal angular momentum.

The density continues to drop as we get closer to the surface of the neutron star. Eventually, individual nuclei can begin to condense out of the degenerate neutron fluid forming the inner crust. This inner crust consists of a mixture of free neutrons, neutron-rich nuclei, and relativistic free electrons. The inner crust is a transition zone where relativistic degenerate electrons supply the pressure and it becomes energetically favorable of some neutrons to exist outside the nuclei. Ordinarily, free neutrons will spontaneously decay into a proton and an electron in a process known as *beta decay*. However, since there are no free electron states in the degenerate gas available for the emitted electrons, the free neutrons remain. The process by which free neutrons escape from neutron-rich nuclei is known as *neutron drip*.

The outer boundary of the inner crust is where the density becomes too low for neutron drip to occur. Above this zone, the outer crust consists of a lattice of nuclei and a sea of relativistic degenerate electrons. Near the surface, the nuclei are mostly ^{56}Fe, but at increasing depths, the Fermi energy of the degenerate electrons becomes high enough that inverse beta decay can occur and the effect is to combine electrons with protons to produce excess neutrons in the nuclei. This process is known as *neutronization*. The nuclei become increasingly neutron-rich, producing elements such as ^{62}Ni, ^{80}Zn, or ^{118}Kr. The relativistic degenerate electrons in the outer crust would normally produce a high conductivity, but the existence of strong magnetic fields within the core inhibits the conductivity in directions orthogonal to

Fig. 12.3 A cartoon of the structure of the interior of a neutron star. The bulk of the neutron star is made up of degenerate neutrons that supply the pressure. The crusts are supported by degenerate electron pressure. The outer crust consists of a lattice of neutron-rich nuclei. The inner crust contains both nuclei and free neutrons as neutron drip begins to be important

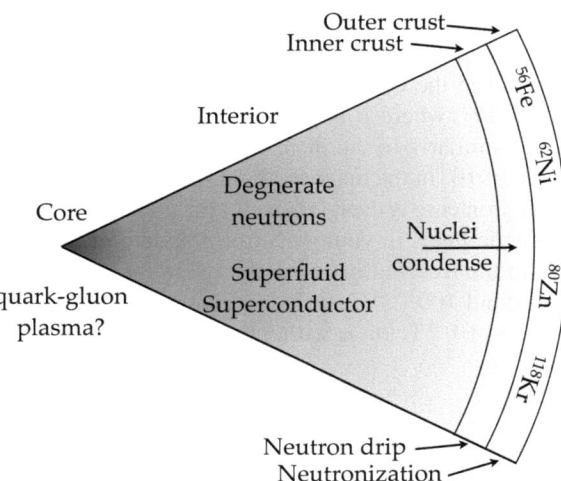

the magnetic field lines crossing the crust. The generic structure of a neutron star is shown in Fig. 12.3. The outer atmosphere of a neutron star is also strongly affected by the magnetic field as we will see in the next section.

The birth of a neutron star is accompanied by the catastrophic collapse of the iron core, followed by a type II supernova explosion. During the collapse of the iron core, it shrinks from a roughly earth-sized (10^4 km) down to about 10 km. The collapse is rapid and the core decouples from the envelope, so the angular momentum of the core is conserved during the collapse. Therefore, the final rotational period, P_f, is related to the initial rotational period, P_0, by

$$P_f = P_0 \left(\frac{R_f}{R_0}\right)^2, \qquad (12.30)$$

where R_0 and R_f are the initial and final radii. The initial rotation of the iron core is not well known, but we can approximate it by noting that the last few stages of nuclear burning take place on very short timescales and are accompanied by contraction at each burning cycle. Thus, starting with a core radius of $\sim 0.3\,R_\odot$ and a typical solar rotation period of ~ 10 days, we find the rotation period of the iron core to be approximately 1 day. After collapse to a neutron star, the rotation period shrinks to ~ 100 ms. The angular momentum of the initial rotation is then frozen into the superfluid core of the neutron star.

Problem 12.2: The minimum rotational period of an object held together by gravity is found by equating the tangential velocity at the equator to the orbital velocity of a test particle at the surface. Assume a neutron star remains spherical even at these high rotation rates and determine the minimum orbital period.

At the same time as the angular momentum is being frozen into the neutron star, any magnetic fields present in the iron core will also be trapped by the superconducting material in the core. The magnetic flux through the surface of the core scales as $\Phi_m \propto BR^2$, where B is the magnetic field. Thus, the growth of the magnetic field scales similarly to the increase in the rotational period, resulting in a growth of a factor of 10^6 in the magnetic field strength. The initial strength of the magnetic field in the iron core is even less well known than the rotational period. However, typical magnetic fields in young neutron stars are on the order of 10^8 T. If this is entirely due to the freezing in of the flux, then the typical iron core magnetic fields would be around 100 T. This value is enormous compared to the solar magnetic field of ~ 1 G $\sim 10^{-4}$ T, but is within the range of typical white dwarf magnetic fields.

12.3 Pulsars

When they are born, neutron stars are very hot, with internal temperatures of $\sim 10^{11}$ K, but they cool down rapidly due to a neutrino emission process whereby a neutron undergoes beta decay to form a proton and an electron, which then recombine through neutralization to form a neutron. The net result is that a neutron emits a neutrino/antineutrino ($\nu \bar{\nu}$) pair that both escape the neutron star and carry away energy:

$$n \to p + e + \bar{\nu} \quad p + e \to n + \nu. \tag{12.31}$$

This process is known as the *URCA process*. It continues to cool the neutron star until the temperature is low enough that there are too few nondegenerate neutrons and protons for there to be available states for the decay products. Typically this is about 10^9 K. Further neutrino-based cooling can occur with neutron collisions emitting $\nu \bar{\nu}$ pairs as the very high-energy neutrons settle down into degenerate states. Combined with photon emission from the surface, the neutron star cools to about 10^6 K over the next few thousand years. Even at these temperatures, the luminosity of neutron stars is less than solar luminosity, and the spectrum peaks in the X-ray band. Thus, neutron stars are quite difficult to observe without X-ray detectors.

In addition to the isotropic thermal emission, neutron stars also emit beamed radiation due to their rapid rotation and the huge magnetic fields that are frozen in at their birth. The radiation is beamed along an axis defined by the dipole moment of the neutron star magnetic field. If the dipole moment vector is not aligned with the angular momentum vector, then this beam of radiation is swept around with the rotational period of the neutron star. If the beam crosses the line of sight to the earth, we see a pulse of radiation. Typically, the radiation is in the radio band of the spectrum and so these objects are given the designation PSR for *pulsating source of radio*. They are called pulsars.

12.3 Pulsars

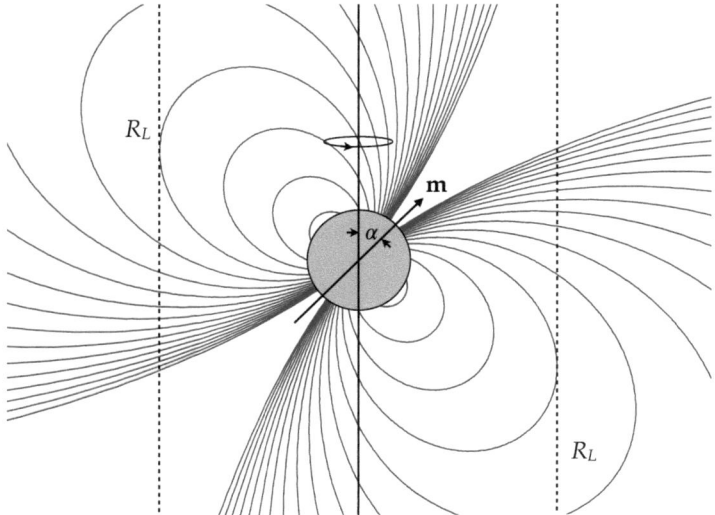

Fig. 12.4 The magnetic field surrounding a neutron star. The dipole magnetic moment **m** is offset by an angle α from the axis of rotation. The lines labeled R_L denote the edge of the light cylinder, where a co-rotating particle would have a tangential velocity equal to c. Outside of the light cylinder, the dipole field lines do not accurately represent the true field lines of the pulsar

Consider the magnetic field to be pure dipole inclined at an angle α with respect to the spin axis, as shown in Fig. 12.4. As the neutron star rotates, the magnetic field near the surface varies rapidly. From Faraday's law, this induces a large electric field near the surface of the neutron star. The electric field dominates even the strong gravity found at the surface and can pull ions and electrons off of the star. Thus, the space surrounding the neutron star is filled with a plasma made up of ions and electrons torn from the surface. This plasma creates a *magnetosphere* of charged particles spiraling along the magnetic field lines and being dragged along with the neutron star's rotation. At a certain distance from the axis of rotation of the neutron star, the tangential velocity of corotating particles will exceed the speed of light. This defines a *light cylinder* aligned with the spin axis and with a radius of:

$$R_L = \frac{c}{\omega} = \frac{cP}{2\pi}, \tag{12.32}$$

where $\omega = 2\pi/P$ for a pulsar with rotational period P. For a typical pulsar with $P = 100$ ms, $R_L = 4.8 \times 10^6$ m, which is about two-thirds of the earth's radius, but about 500 times larger than the neutron star itself. At this radius, the plasma can no longer keep up with the rotation and it distorts the magnetic field beyond the light cylinder, carrying it away in a pulsar wind. Plasma near the poles of the magnetic field are strongly accelerated and generate both *curvature* radiation and *synchrotron* radiation which after some reprocessing through the plasma eventually become the radio pulses observed at large distances.

The main source of energy for the pulses and pulsar wind is the kinetic energy of rotation of the neutron star. Although the interaction between the dipole magnetic field and the plasma of the magnetosphere prevents the magnetic field from radiating a pure dipole radiation field, we can estimate the rate at which rotational kinetic energy is lost by looking at the power radiated by a rotating magnetic dipole and assuming that this energy is simply reprocessed into the pulsar wind and radio pulses. If we define a Cartesian coordinate system with the z-axis along the spin axis of the pulsar, the magnetic dipole moment shown in Fig. 12.4 is given by

$$\mathbf{m} = \frac{\sqrt{\pi}}{\mu_0} BR^3 \left(\mathbf{e}_x \sin\alpha \cos\omega t + \mathbf{e}_y \sin\alpha \sin\omega t + \mathbf{e}_z \cos\alpha \right). \quad (12.33)$$

A magnetic dipole moment radiates power at a rate related to the second time derivative of the magnetic moment. Thus, the radiated power is

$$L = \frac{2}{3c^3} |\mathbf{\ddot{m}}|^2 = \frac{4\pi B^2 R^6 \omega^4 \sin^2\alpha}{6c^3 \mu_0}. \quad (12.34)$$

We assume this power is reprocessed into pulsar radiation and the pulsar wind through the plasma in the magnetosphere. The energy to drive this power comes from the rotational kinetic energy; thus the pulsar should spin down as this energy is lost. Relating the time derivative of the rotational kinetic energy to the radiated power gives up a relation between the period, P, the period derivative, $\dot{P}$, and the magnetic field, B. Thus,

$$I\omega\dot{\omega} = -\frac{2}{3c^3} |\mathbf{\ddot{m}}|^2 = -\frac{4\pi B^2 R^6 \omega^4 \sin^2\alpha}{6c^3 \mu_0} \quad (12.35)$$

or

$$\dot{P} = \frac{8\pi^3 B^2 R^6 \sin^2\alpha}{3c^3 \mu_0 PI}, \quad (12.36)$$

where I is the moment of inertia of the neutron star.

Problem 12.3: Consider a pulsar with a mass of $1.4 M_\odot$ and a magnetic field of $B = 10^8$ T at an angle of $\alpha = 30°$.

(a) Assume that the pulsar has constant density and calculate its moment of inertia.
(b) Define the characteristic lifetime of a pulsar to be $\tau = P/\dot{P}$ and compute the characteristic lifetimes for $P = 100$ ms, $P = 10$ ms, and for the minimum period.

Pulsars are very precise clocks and both the pulse period and its time derivative can be measured very accurately. Plotting $\dot{P}$ versus P shows that most pulsars are born with short orbital periods and high magnetic fields which both gradually decay as the pulsar loses energy to the wind and pulse emission. One class of pulsars that show up in a P-$\dot{P}$ plot have short orbital periods, but low values of $\dot{P}$. From Eq. (12.36), we see that this implies a low magnetic field strength. These systems are assumed to have been recycled by being spunup through the accretion of matter from a companion.

12.4 Black Holes

If the mass of the central core is large enough at the end of a massive star's life, then degenerate neutron pressure will be unable to halt the collapse. There is no other source of pressure known that will prevent the radius of the core from shrinking beyond the point at which the escape velocity equals the speed of light. Clearly, at this point, relativity is necessary to describe these compact objects. Although a detailed treatment using general relativity is beyond the scope of this book, several general features can be determined using a combination of special relativity and Newtonian gravity. The critical surface at which $v_{esc} = c$ is called the "event horizon." Below this surface no information can escape from the black hole and so the event horizon is frequently used to describe the size of the black hole although it is not a solid surface at all. For a spherically symmetric black hole, the event horizon is described by the *Schwarzschild radius*. Using Newtonian gravity, the Schwarzschild radius would be found from equating the escape velocity to c:

$$v_{esc} = c = \sqrt{\frac{2GM}{R_s}} \implies R_s = \frac{2GM}{c^2}. \qquad (12.37)$$

A detailed solution of the vacuum equations for general relativity yields precisely the same result.

Given that the speed of light should be constant, one might ask how light behaves when the escape velocity is equal to the speed of light. To fully answer this question, we must use general relativity, which describes the gravitational field around an object as a manifestation of curved space-time. We can gain some understanding of the effects of a curved space-time by looking at the *metric*, which is important in defining distances in space-time that are *invariant* or independent of coordinate choice. Around a nonrotating, spherically symmetric black hole, the invariant infinitesimal distance is described by the *Schwarzschild* metric, so

$$ds^2 = \left(1 - \frac{R_s}{r}\right) c^2 dt^2 - \left(1 - \frac{R_s}{r}\right)^{-1} dr^2 - r^2 \left(d\theta^2 + \sin^2\theta d\phi^2\right), \qquad (12.38)$$

where the coordinates r, θ, and ϕ are closely related (but not identical) to the usual spherical polar coordinates and t is the time measured by a clock at infinity. The proper time ($d\tau$) measured at radius r from the black hole is found by setting all the spatial intervals to zero, so

$$d\tau = dt\sqrt{1 - \frac{R_s}{r}}. \tag{12.39}$$

If we consider $d\tau$ to measure the local period of oscillation of an electromagnetic field at r, then the frequency of light emitted at radius r is redshifted by an amount given by

$$f' = f_0\sqrt{1 - \frac{R_s}{r}} \tag{12.40}$$

for an observer at infinity. Note that at $r = R_s$, the frequency is redshifted to zero, and so the photon carries no energy.

As we have seen with neutron stars, the remnant core of a massive star will carry some angular momentum with it, and so it spins up as it collapses. Therefore, we should expect black holes to also possess angular momentum and so the Schwarzschild metric is inadequate for describing real black holes. The appropriate metric for a spinning black hole with an angular momentum J, is the *Kerr* metric. In a coordinate system which approximates spherical polar coordinates at large distances, the invariant infinitesimal distance described by the Kerr metric is

$$ds^2 = \left(1 - \frac{R_s r}{\rho^2}\right)c^2 dt^2 - \frac{\rho^2}{\Delta}dr^2 - \rho^2 d\theta^2 - \left(r^2 + \alpha^2 + \frac{R_s r \alpha^2}{\rho^2}\sin^2\theta\right)\sin^2\theta \, d\phi^2$$
$$+ \frac{2R_s r\alpha \sin^2\theta}{\rho^2} c \, dt \, d\phi, \tag{12.41}$$

where R_s is still the Schwarzschild radius and

$$\alpha = \frac{J}{Mc} \tag{12.42}$$

$$\rho^2 = r^2 + \alpha^2 \cos^2\theta, \tag{12.43}$$

$$\Delta = r^2 - R_s r + \alpha^2. \tag{12.44}$$

A couple of notable features about the line element in the Kerr metric are that (1) there is a cross term involving $dt d\phi$ and (2) the coefficients in front of dt^2 and dr^2 are no longer reciprocals of each other.

There are two important surfaces around spinning black hole. When $\Delta = 0$, the coefficient in front of dr^2 becomes undefined. This surface gives the event horizon at which it is impossible to escape from the black hole. Clearly the solution to $\Delta = 0$ gives two values for r. The outer value is the event horizon:

12.4 Black Holes

Fig. 12.5 The basic shape of the event horizon and ergosphere for a spinning black hole

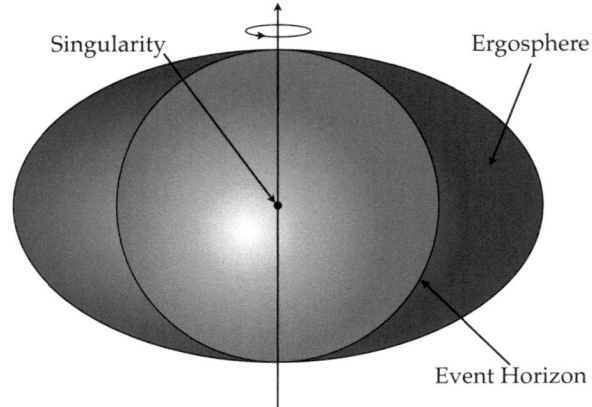

$$R_h = \frac{R_s}{2}\left(1 + \sqrt{1 - \frac{4\alpha^2}{R_s^2}}\right). \quad (12.45)$$

Note that $R_h \leq R_s$ for all values of α, so that in some sense spinning black holes are smaller than Schwarzschild black holes. The second surface is defined as the radius at which the coefficient in front of $c^2 dt^2$ goes to zero. It is still possible to escape from the black hole from points between this surface and the event horizon. If we assume that $dr = d\theta = d\phi = 0$, then the space-time interval becomes zero at this surface. A space-time interval of zero implies that the path is followed by a particle traveling at the speed of light. Therefore, just below this surface, it is impossible to remain at rest with respect to the distant stars. This means that the space-time itself is being dragged around by the spin of the black hole. This space within this surface is called the *ergosphere*. Again, there are two solutions for the ergosphere, but the inner one is within the event horizon, so the outer one defines the ergosphere:

$$R_e = \frac{R_s}{2}\left(1 + \sqrt{1 - \frac{4\alpha^2}{R_s^2}\cos^2\theta}\right). \quad (12.46)$$

The horizon structure of a spinning black hole is shown in Fig. 12.5.

Problem 12.4: Determine the maximum value of the angular momentum that a black hole may have and still have an event horizon.

An isolated black hole is clearly invisible, but it can make itself known in the presence of a companion. At large separations, a black hole can be inferred by its gravitational influence on a visible companion in a binary system. At closer separations, the black hole can draw matter off of its companion. The infalling material must lose energy as it is accreted into the black hole. This energy is radiated away as X-rays.

Problems

12.1. Calculate the effective temperature and the core temperature of a $0.7\,M_\odot$ white dwarf if its luminosity is $10^{-3}\,L_\odot$.

12.2. The minimum rotational period of an object held together by gravity is found by equating the tangential velocity at the equator to the orbital velocity of a test particle at the surface. Assume a neutron star remains spherical even at these high rotation rates and determine the minimum orbital period.

12.3. Consider a pulsar with a mass of $1.4\,M_\odot$ and a magnetic field of $B = 10^8\,\text{T}$ at an angle of $\alpha = 30°$.

(a) Assume that the pulsar has constant density and calculate its moment of inertia.
(b) Define the characteristic lifetime of a pulsar to be $\tau = P/\dot{P}$ and compute the characteristic lifetimes for $P = 100\,\text{ms}$, $P = 10\,\text{ms}$, and for the minimum period.

12.4. Determine the maximum value of the angular momentum that a black hole may have and still have an event horizon.

Part IV
Dynamical Systems

Up to half of all stars in the sky are thought to be binary systems. This implies that nearly two-thirds of all stars evolve in the presence of a companion that may alter their evolution. Furthermore, many stars are born in clusters. If the clusters are dense enough, the dynamics of the stars in the cluster can cause single and binary stars to encounter one another during their lives. These encounters can affect the further evolution of the stars involved. Here we examine the consequences of dynamical processes on the evolution of stars.

Chapter 13
Binary Evolution

Stars in binary systems do not evolve in isolation. If the orbital separation is small enough, the tidal perturbations due to the companion can break the spherical symmetry that was assumed in the development of the evolution equations in Chap. 4 and then used in later chapters. For the most part, these additional considerations do not significantly alter the evolution of the components of a binary since the orbital periods are generally greater than dynamical timescales, yet much shorter than thermal timescales. However, if the stars begin to directly interact through mass transfer and mass loss, then the bulk properties of the components can change in mid-evolution, resulting in a dramatically different outcome in the evolution compared to isolated stars with the same properties. In this chapter, we will look at the physics of mass transfer and the consequences of mass transfer and mass loss on the evolution of the components of a binary as well as the binary system itself.

13.1 The Roche Model

We are interested in understanding the gravitational potential about a binary system so that we can determine the motion of test bodies under the influence of the gravitational field of both components of the binary. The behavior of the matter in both stars will be governed by this combined potential. For simplicity, we will consider only circular binaries in this section. In effect, we are looking at a three-body problem. We consider the mass of the third body, m, to be infinitesimally small compared with the components of the binary, so that its presence does not affect the motion of the binary. In this case we can choose coordinates that rotate with the binary, so that the components of the binary are at fixed coordinate positions. We choose them to lie along the y-axis in this coordinate system with the origin located at the center of mass of the system. We place the more massive star, m_1, a distance a_1 in the negative y-direction, and the less massive star m_2 is a distance a_2 in the positive y-direction. The rotational frequency is then

Fig. 13.1 Configuration of the three-body problem with the third body as a test mass using corotating coordinates. The origin is placed at the center of mass of the system and the rotation rate is determined by the orbital frequency of the binary system consisting of m_1 and m_2

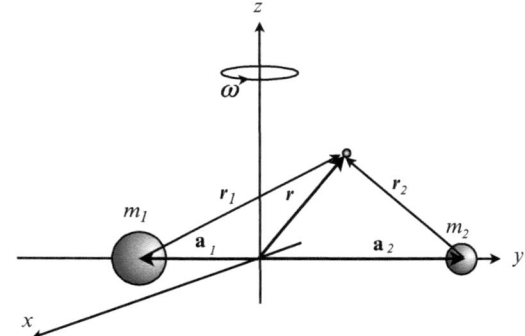

$$\omega = \frac{2\pi}{P} = \sqrt{\frac{G(m_1+m_2)}{a^3}}, \quad (13.1)$$

with $a = a_1 + a_2$. The configuration is shown in Fig. 13.1. We can now write a Lagrangian for the third body in these coordinates, making sure to include the kinetic energy of rotation about the z-axis:

$$\mathscr{L} = \frac{1}{2}m\left(\dot{x}^2+\dot{y}^2+\dot{z}^2\right) + \frac{1}{2}m\omega^2\left(x^2+y^2\right) + \frac{Gm_1m}{r_1} + \frac{Gm_2m}{r_2}, \quad (13.2)$$

where $r_1 = |\mathbf{r} - \mathbf{a}_1|$ and $r_2 = |\mathbf{r} - \mathbf{a}_2|$. Note that we can group the last three terms of the Lagrangian as a *pseudopotential* in the non-inertial corotating frame and the test mass will then act like a particle moving under the influence of this pseudopotential:

$$\Phi' = \frac{1}{2}m\omega^2\left(x^2+y^2\right) + \frac{Gm_1m}{r_1} + \frac{Gm_2m}{r_2}. \quad (13.3)$$

This has the effect of adding a fictitious "centrifugal" force to the problem. When plotting out the equipotential surfaces of the pseudopotential, it is convenient to translate the coordinates so that the origin lies on top of m_1 and m_2 is then located at $y = a$. We can then rescale our lengths so that $a = 1$ and

$$a_1 = \frac{m_2}{m_1+m_2}, \quad (13.4)$$

and $r_1 = \sqrt{x^2+y^2}$ and $r_2 = \sqrt{x^2+(y-1)^2}$. In the translated coordinates, the pseudopotential becomes

$$\Phi = -\frac{Gm_1}{r_1} - \frac{Gm_2}{r_2} - \frac{\omega^2}{2}\left[x^2 + \left(y - \frac{m_2}{m_1+m_2}\right)^2\right]. \quad (13.5)$$

We define $\Phi_n = -2\Phi/G(m_1+m_2)$ to be the *normalized potential* and define the mass ratio to be $q = m_2/m_1$ so that $0 \leq q \leq 1$. Then, we have

$$\Phi_n = \frac{2}{(1+q)r_1} + \frac{2}{(1+q)r_2} + x^2 + \left(y - \frac{q}{(1+q)}\right)^2. \quad (13.6)$$

13.1 The Roche Model

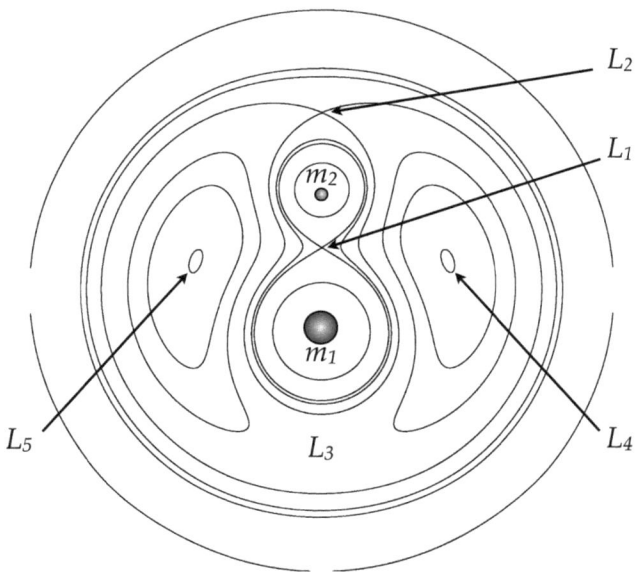

Fig. 13.2 Cross section of equipotential surfaces in the orbital plane of a binary with $q = 0.4$

Equipotential surfaces can be found by setting $\Phi_n = $ constant. A cross section of equipotential surfaces in the xy-plane is shown in Fig. 13.2. All distances are in units of a, and the shape of the surfaces is dependent upon q. There are five extrema in the orbital plane, known as *Lagrange points*. A point between the two masses along the line joining them is the saddle point called L_1. The equipotential surface containing L_1 outlines the *Roche lobes* for both stars. Particles within the Roche lobes are under the gravitational control of a single star. The L_1 point is also referred to as the inner Lagrange point. There is an outer Lagrange point (L_2) located at $y > 1$ and a slightly higher potential point at L_3 at $y < 0$. There are two maxima at the third vertex of equilateral triangles with m_1 and m_2 at the other two vertices. Remember that this surface is in a corotating frame, and so these points are also rotating about the center of mass of the system.

Very close to each component, the equipotential surfaces are nearly spherical and therefore stars that are small compared to their Roche lobes evolve as if they were in isolation. However, as the stars evolve, their radii increase. If the radius of a star increases to the point that it begins to approach its Roche lobe, then the shape of the star begins to become nonspherical as its surface is defined by the equipotential surface which is also nonspherical. Although stellar evolution generally assumes spherical symmetry, we usually work with locally defined quantities such as density. Therefore, we are more concerned with the volume of the star, rather than its radius. Hence, we introduce the effective radius, r_L, of the Roche lobe. This is the radius of a sphere that has the same volume as the Roche lobe. There are several approximations used for r_L based on numerical modeling of the Roche geometry. The most widely used one is from Eggleton and is calculated from the normalized potential to be

$$r_L = \frac{0.49 q^{2/3}}{0.69 q^{2/3} + \ln(1 + q^{1/3})}, \quad (13.7)$$

where r_L is the effective Roche lobe radius about the star whose mass is in the numerator of q. Thus, the value of r_{L2} is computed using $q = m_2/m_1$, while the value of r_{L1} requires using $q = m_1/m_2$. Since these radii are based on the normalized potential, the actual *Roche lobe radius* is

$$R_L = r_L a. \quad (13.8)$$

Once a star fills its Roche lobe, mass transfer begins as matter from the Roche lobe filling star can now flow across L_1 in a process known as *Roche lobe overflow* or RLOF. During mass transfer, the mass-losing star is called the *donor* and the other star is called the *accretor*. The transfer of matter between stars changes the mass ratio of the system and therefore the Roche lobe geometry changes. In addition, the loss of mass from the Roche lobe filling star can alter the rate at which its radius changes. If the relative rate of change between the Roche lobe and the Roche lobe filling star is such that it drives the surface of the star even further beyond the Roche lobe, then the mass transfer will become unstable.

13.2 Mass Transfer Stability

We will first consider the case of conservative mass transfer in which no mass is lost from the system. In this case, angular momentum is conserved and $\dot{m}_1 = -\dot{m}_2$. Using Eq. (2.71) for the angular momentum of a circularized binary, we have

$$J = m_1 m_2 \sqrt{\frac{Ga}{M}}, \quad (13.9)$$

where $M = m_1 + m_2$ is the total mass of the system. Since $\dot{J} = 0$, the rate of change of the orbital separation can be expressed in terms of the rate of change of the primary mass:

$$\dot{a} = \frac{2(m_1 - m_2)a}{m_1 m_2} \dot{m}_1. \quad (13.10)$$

Note that, since $m_1 \geq m_2$, if the more massive star is the donor, then the orbit shrinks, and if the less massive star is the donor, then the orbit expands.

If we want to know the rate at which the Roche radius changes with mass transfer, we need to write Eq. (13.7) in a way that includes the change in a. Note that the orbital separation can also be written as

$$a = \frac{J^2}{GM^3} \frac{(1+q)^4}{q^2} = a_c \frac{(1+q)^4}{q^2}, \quad (13.11)$$

13.2 Mass Transfer Stability

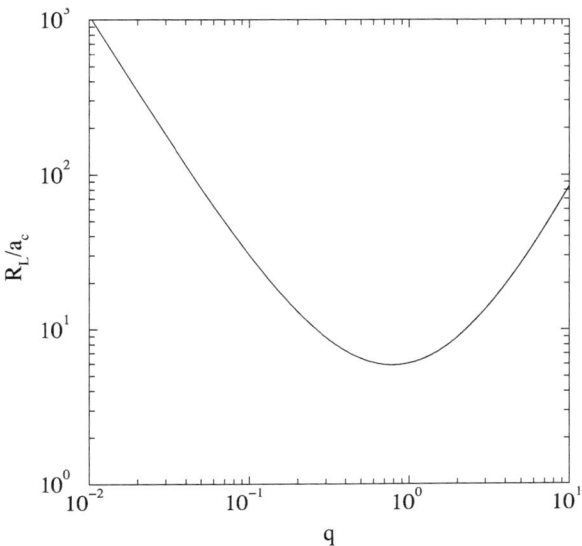

Fig. 13.3 The Roche lobe radius in units of a_c as a function of q. Note that the slope is -2 for small q and 2 for large q. The minimum occurs near $q = 1$

where $a_c = J^2/GM^3$ is a constant throughout the mass transfer. The Roche lobe radius is then

$$R_L = a_c \frac{0.49\,(1+q)^4}{0.69q^2 + q^{4/3}\ln\left(1+q^{1/3}\right)}. \tag{13.12}$$

When both components of the binary are main sequence stars, the more massive star usually evolves off the main sequence and fills its Roche lobe first. Therefore, we should consider $q = m_1/m_2 > 1$ in Eq. (13.12). During mass transfer, it is quite possible that m_2 can gain enough matter from m_1 so that it becomes the more massive star. Thus, it is important to look at the behavior of R_L for the full range of q (i.e., $0 < q < \infty$). At small values of q, $R_l \propto q^{-2}$, while at large values of q, $R_L \propto q^2$. The minimum lies around $q \simeq 0.8$. Figure 13.3 shows R_L/a_c for $0.01 < q < 10$.

The rate at which a radius changes with mass is usually described by the *radius–mass exponent*, $\zeta = d\ln R/d\ln M$, which is the logarithmic derivative of the radius with respect to the mass. For the Roche lobe radius–mass exponent, we want to know how R_{L1} varies with m_1, on the assumption that m_1 is the donor. Since Eq. (13.12) gives R_L as a function of q, we note that

$$\frac{\partial \ln R_{L1}}{\partial \ln m_1} = \frac{m_1}{R_{L1}} \frac{\partial q}{\partial m_1} \frac{\partial R_{L1}}{\partial q}. \tag{13.13}$$

Because the mass transfer is conservative, m_2 also varies as m_1 is changing. Therefore we express m_1 as a function of M and q:

$$m_1 = M \frac{q}{1+q}. \tag{13.14}$$

Thus,
$$\zeta_{L1} = \frac{\partial \ln R_{L1}}{\partial \ln m_1} = (1+q)\frac{\partial \ln R_{L1}}{\partial \ln q}. \tag{13.15}$$

From here, we find that the radius–mass exponent for the Roche lobe is

$$\zeta_{L1} = 4q - \frac{2}{3}(1+q)\left[2 + \frac{0.6q^{2/3} + 0.5q^{1/3}\left(1+q^{1/3}\right)^{-1}}{0.6q^{2/3} + \ln\left(1+q^{1/3}\right)}\right]. \tag{13.16}$$

Although it is not immediately apparent from the equation, ζ_{L1} is a monotonically increasing function of q. In the limit $q \to 0$, $\zeta_{L1} \to -5/3$, while $\zeta_{L1} \to 2q$ for large q. In fact, it can be closely approximated by a line.

Mass transfer begins when one component of the binary fills its Roche lobe. Since the mass transfer itself will cause q to change, the Roche radius will also change. Similarly, mass loss from the donor star will cause its radius to change. The stability of mass transfer will depend on the relative rates of change between the stellar radius and the Roche radius. If the mass transfer drives the Roche radius to become smaller than the stellar radius, then a runaway scenario will ensue and the mass transfer will become unstable. On the other hand, if the mass transfer tends to expand the Roche radius relative to the donor radius, then the mass transfer is stable. Consequently, the radius–mass exponent of the donor star, combined with the mass ratio of the binary, determines the stability of the mass transfer episode.

The radius–mass exponent for the mass donor depends on both the mass of the donor as well as its evolutionary state. Generally, if the donor star is not degenerate, then the transfer event will occur while the star is either burning hydrogen on the main sequence, expanding to a red giant, or burning helium. These three states are classified as *case A*, *case B*, or *case C* mass transfer. The typical radius of a $10 M_\odot$ star is shown in Fig. 13.4. At the onset of mass transfer, the star can only adjust adiabatically as the mass transfer occurs on timescales much shorter than the thermodynamic timescale. We compute the radius–mass exponent for the donor, ζ_* in the adiabatic limit. At the onset of mass transfer, $R = R_{L1}$. If $\zeta_* > \zeta_L$, then the radius of the star becomes smaller than the Roche radius and the rate of mass transfer is driven by stellar evolution causing the radius of the star to increase in size at constant mass. Stable mass transfer therefore occurs on either a thermodynamic or nuclear timescale. On the other hand, if $\zeta_* < \zeta_L$, then the radius of the star grows larger than the Roche radius and the mass transfer is unstable. Unstable mass transfer occurs on a dynamical timescale. There is a critical value of mass ratio, q_c, when $\zeta_* = \zeta_L$ for a given type of star. Because ζ_L is a monotonic function of q, binaries with $q > q_c$ will be unstable to mass transfer.

Polytropic models with adiabatic index $\gamma_a = 5/3$ are good approximations to stars with convective envelopes. From Eq. (8.34)

$$R^{3-n} \propto m^{1-n}, \tag{13.17}$$

13.2 Mass Transfer Stability

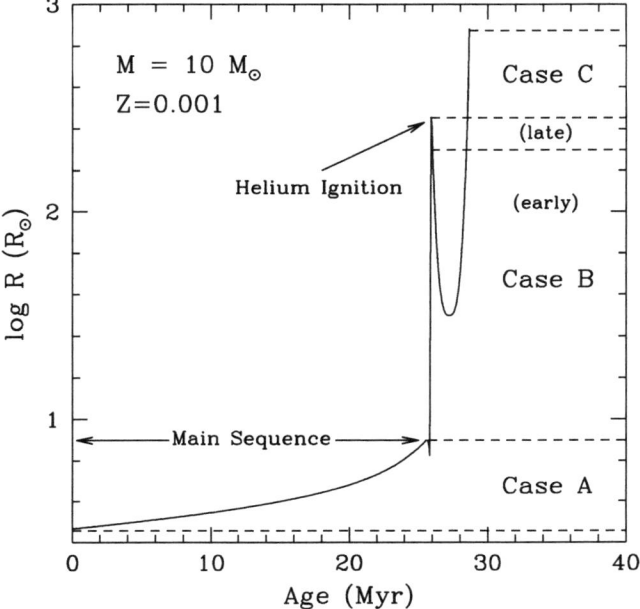

Fig. 13.4 Evolution of the radius of a $10 M_\odot$ star with a metallicity of $Z = 0.001$, showing the times of case A, case B, and case C mass transfer. Figure from Pfahl, Rappaport, & Podsiadlowski, *Astrophysical Journal* 573, 283 (2002). Reproduced by permission of the AAS

we see that

$$\zeta_* = \frac{1-n}{3-n}. \tag{13.18}$$

Therefore, for $n = 1.5$, $\zeta_* = -1/3$ for stars with convective envelopes. For fully radiative envelopes, we would expect $\gamma_a = 4/3$, and so the polytropic index is $n = 3$. Unfortunately, Eq. (8.34) then gives $\zeta_* = -2/0$, which is undefined. We are then forced to use more sophisticated models to determine ζ_*. Typical results for main sequence stars with $M < M_\odot$ give $\zeta_* \sim 0.7$. For more massive stars with radiative envelopes, then $\zeta_* > 2$. These typical values are shown in Fig. 13.5, along with the curve of $\zeta_L(q)$. It is interesting to note that the radius–mass exponent for degenerate objects such as white dwarfs is also $\zeta_{wd} = -1/3$.

During mass transfer, the companion to the donor is accreting matter that is falling into its Roche lobe from the donor. The matter entering from the L_1 point must lose its energy in order to fall into the companion. This causes the matter to heat up and radiate away the lost energy. The luminosity of accretion can be estimated by assuming the matter is falling in from infinity and so

$$L_{\text{acc}} = \frac{G m_2 \dot{m}_2}{R_2}. \tag{13.19}$$

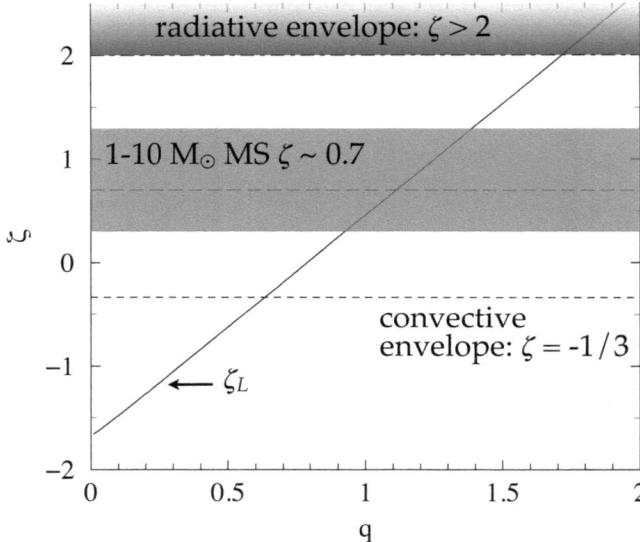

Fig. 13.5 The curve of ζ_L as a function of q. Typical values of ζ_* are given as *horizontal lines* in this plot. The value of q_c is found by setting $\zeta_L = \zeta_*$. For values of $q > q_c$, mass transfer is dynamically unstable

where we have assumed that $\dot{m}_2$ is positive. If the luminosity of accretion is greater than the Eddington luminosity, then radiation pressure will be strong enough to prevent the matter from accreting. This sets an upper bound on the accretion rate, found from

$$L_{\rm acc} < L_{\rm Edd} = \frac{4\pi c G m_2}{\kappa}, \tag{13.20}$$

where κ is usually taken to be the value for electron scattering:

$$\kappa_{\rm es} = 0.02\,(1+X)\,{\rm m}^2/{\rm kg}. \tag{13.21}$$

Assuming the accreted matter is all hydrogen, then $\kappa_{\rm es} = 0.04\,{\rm m}^2/{\rm kg}$, and the Eddington accretion rate is

$$\dot{M}_{\rm Edd} = 3.3 \times 10^{-4}\,{\rm M}_\odot/{\rm year}\left(\frac{R}{{\rm R}_\odot}\right). \tag{13.22}$$

13.3 Unstable Mass Transfer and Mass Loss

There are two basic ways that mass transfer can be nonconservative. If the mass transfer is dynamically unstable, then the radius of the donor star continues to expand beyond the Roche lobe. In this way, the envelope of the donor star grows to encompass the companion and a *common envelope* is formed around both stars. If the mass transfer is dynamically stable, but matter is transferred at a greater rate than the companion can accrete it, then the transferred matter also eventually accumulates into a common envelope about both stars. When a common envelope is formed, then orbital motion of the companion and the core of the donor star can mechanically eject the common envelope. The energy required to unbind the envelope from the binary will come from the orbital energy of the system. Thus, the core and the companion will end up closer together. The details of this process are still not well modeled, but two prescriptions for determining the final orbital state of the binary are commonly used.

The most commonly used approach is the α-*prescription* in which the energy required to eject the envelope comes from the orbital energy of the binary and thus the orbit shrinks. The efficiency of this process determines the final orbital period after the common envelope is ejected. The efficiency parameter is defined as

$$\alpha = \frac{\Delta E_{\text{bind}}}{\Delta E_{\text{orb}}}, \tag{13.23}$$

where ΔE_{bind} is the binding energy of the envelope just prior to mass transfer and ΔE_{orb} is the change in orbital energy during the ejection of the common envelope. The efficiency parameter can be thought of as the fraction of orbital energy that goes into unbinding the envelope. The binding energy of the envelope is

$$\Delta E_{\text{bind}} = -\frac{G m_1 m_1^e}{\lambda R_{L1}}, \tag{13.24}$$

where m_1 is the initial mass of the donor, m_1^e is the mass of the envelope, R_{L1} is the Roche lobe radius of the donor, and λ is an averaging factor (of order 1), used to account for the density distribution of the matter in the envelope. The change in the orbital energy of the systems is

$$\Delta E_{\text{orb}} = -\frac{G}{2}\left[\frac{m_1^c m_2}{a_f} - \frac{m_1 m_2}{a_i}\right], \tag{13.25}$$

where m_1^c is the mass of the core of the donor, m_2 is the mass of the accretor, a_f is the final orbital separation, and a_i is the initial orbital separation at the onset of mass transfer. The final orbital separation is obtained from solving Eq. (13.23) for a_f. Typical values of α are on the order of unity.

The other approach is the γ-*prescription*, in which the lost mass carries away angular momentum that comes from the total angular momentum of the binary. The parameter γ describes the fraction of angular momentum carried away by the ejected envelope. Thus,

$$J_f = J_i \left(1 - \gamma \frac{\Delta m}{m}\right), \tag{13.26}$$

where $m = m_1 + m_2$ and $\Delta m = m_1^e$. Again, the final orbital separation is found from solving Eq. (13.26) for J_f. Although this method uses angular momentum loss to determine the final state of the binary, energy is still conserved.

In addition to nonconservative mass transfer, mass can be lost from a binary during catastrophic events. Supernovae can occur during the evolution of a high-mass star. Binary systems can also undergo novae when the accreted hydrogen or helium on the surface of a white dwarf reaches an ignition point and nuclear shell burning occurs on the surface. In these cases, the mass from the explosion is expelled quickly from the binary system. Usually the amount of mass lost during a nova is small compared to the total mass of the system, so the effect is a minor adjustment of the binary system. In a supernova, a large fraction of the mass of the exploding star can be lost. If this star is the more massive star in the binary, the effect on the binary can be quite dramatic.

Let us consider the example of a circular system with semimajor axis a in which a star with mass m_1 explodes leaving behind a compact remnant with mass m_{co}. The compact remnant can be either a black hole or a neutron star. Using the virial theorem for binaries, the initial energy of the binary can be written:

$$E = -\frac{Gm_1 m_2}{2a}. \tag{13.27}$$

After the explosion, the expanding mass shell will quickly cross the orbit of m_2, decreasing the gravitational force acting on the secondary. The new energy of the system will be

$$E' = \frac{1}{2}m_{co}v_1^2 + \frac{1}{2}m_2 v_2^2 - \frac{Gm_{co}m_2}{a}. \tag{13.28}$$

The passage of the mass shell is unlikely to impart significant momentum to the companion, so we can assume that v_2 is unchanged. On the other hand, the explosion is quite likely to impart significant momentum to the remnant, as neutron stars are known to receive kicks at birth resulting in velocities around 200–500 km/s. The direction of the kick may be random, and so it can either increase or decrease the kinetic energy of the remnant. In order to gain some understanding of the outcome of catastrophic mass loss, we will assume that the kick velocity is precisely zero. In this case, the final velocities of both m_2 and m_{co} are unchanged from the initial velocities. Therefore,

$$E' = \frac{1}{2}\left(\frac{m_{co}m_2}{m_{co}+m_2}\right)v^2 - \frac{Gm_{co}m_2}{a}, \tag{13.29}$$

where

$$v^2 = \frac{G(m_1+m_2)}{a}. \tag{13.30}$$

Therefore, the final energy is

$$E' = \frac{Gm_{co}m_2}{2a}\left(\frac{m_1+m_2}{m_{co}+m_2}-2\right). \tag{13.31}$$

If the final energy of the system is positive, then the binary will be disrupted. We can see that this implies that the binary will not survive the loss of more than half of its mass. If we include the possibility of kicks to the compact remnant, then this criterion is not exact, since a kick can add or remove kinetic energy from the system.

If the binary is not disrupted, the new orbit will become eccentric and expand to a larger semimajor axis, given by

$$a' = a\left(\frac{m_2 + m_{co}}{m_2 - m_1 + 2m_{co}}\right), \quad (13.32)$$

with orbital period

$$P' = P\left(\frac{a'}{a}\right)^{3/2}\left(\frac{2a' - a}{a'}\right)^{1/2}. \quad (13.33)$$

Noting that the supernova ejecta also carries away angular momentum given by

$$\Delta J = \frac{2\pi a^2}{P}\frac{(m_1 - m_{co})m_2^2}{(m_1 + m_2)^2} \quad (13.34)$$

we can calculate the eccentricity using

$$J_f = \frac{2\pi a'^2}{P}\frac{m_{co}m_2}{m_{co} + m_2}\sqrt{1 - e^2} = \frac{2\pi a^2}{P}\frac{m_1 m_2}{m_1 + m_2} - \Delta J. \quad (13.35)$$

13.4 Binary Evolution Example

The preceding discussion included a number of simplifications in order to illustrate the basic physics behind mass transfer and binary evolution. Here we give two examples from a full binary stellar evolution code. The first example describes the formation of a *microquasar*, consisting of a black hole accreting matter from main sequence companion, shown in Fig. 13.6. The second example describes the formation of a double neutron star system, shown in Fig. 13.7.

The initial system for the microquasar consists of a primary with ZAMS mass $m_1 = 42.8\,M_\odot$ and a secondary with $m_2 = 1.22\,M_\odot$. The binary is highly eccentric, with $e = 0.6$. The semimajor axis is $a = 5,330\,R_\odot = 24.8\,AU$. After roughly 4.6 Myr have passed, the primary evolves off the main sequence and begins to grow into a giant. When the radius of the giant approaches the size of the periastron, then tidal interactions become efficient and the orbit circularizes. When this happens, energy is removed from the orbit, while the angular momentum is conserved. This results in a circular orbit with a semimajor axis of $a' = a(1 - e^2)$. By the time the primary has grown to fill its Roche lobe, it has lost $17.3\,M_\odot$ to stellar winds, and mass transfer begins for a Roche lobe filling giant star with $m_1 = 25.5\,M_\odot$ in orbit around a $m_2 = 1.22\,M_\odot$ main sequence star. The mass ratio is high enough that the mass transfer is dynamically unstable and a common envelope develops. During the common envelope phase, the primary loses its envelope and its mass drops to

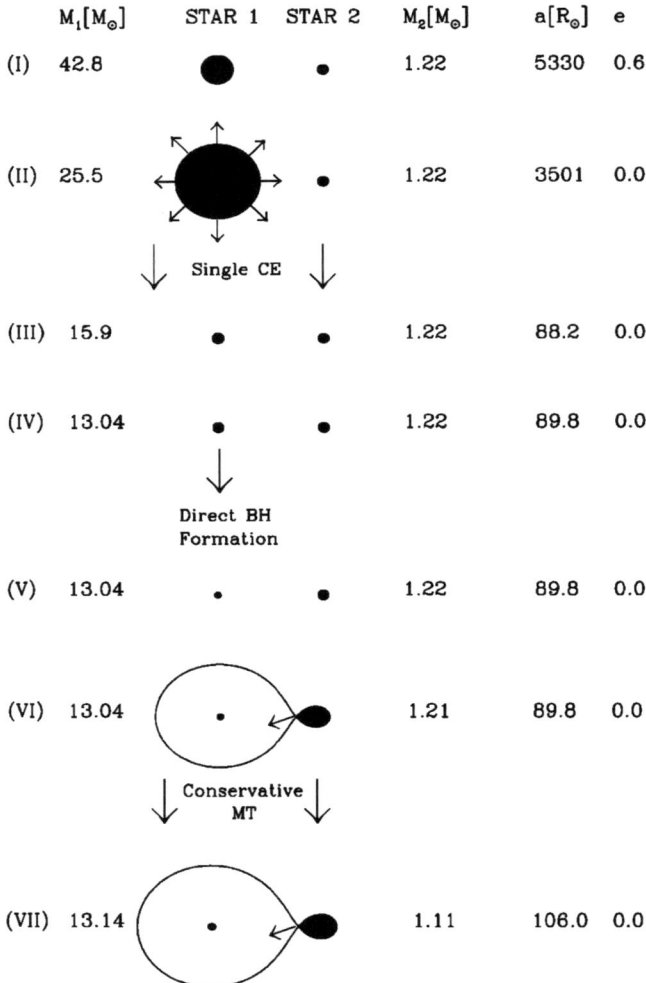

Fig. 13.6 Evolutionary phases leading to a model for the microquasar GRS 1915+105. Figure taken from Belczynski and Bulik *Astrophysical Journal Letters* 574, L147 (2002). Reproduced by permission of the AAS

$15.9\,M_\odot$. The semimajor axis of the orbit shrinks dramatically from $3{,}501\,R_\odot$ down to $88.2\,R_\odot$.

Problem 13.1: Assuming no mass loss from the stellar winds, use conservation of angular momentum to compute the semimajor axis of the tidally circularized orbit for step (I) to (II) in the microquasar example. How does it compare to the value of $3{,}501\,R_\odot$ obtained from the results of the stellar evolution code?

13.4 Binary Evolution Example

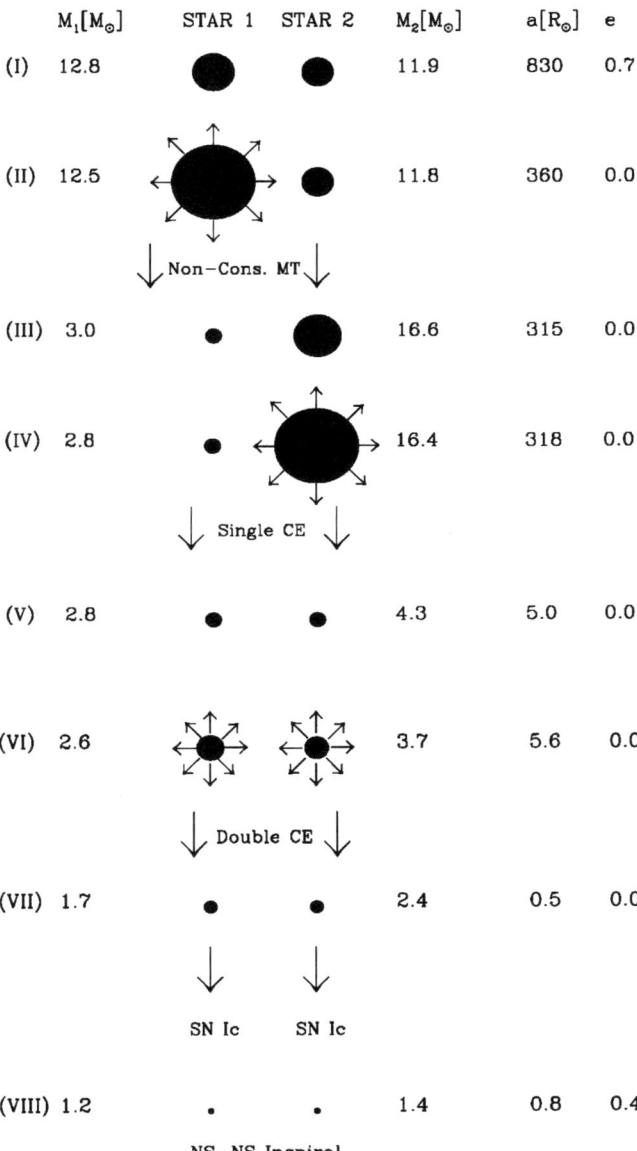

Fig. 13.7 Evolutionary phases leading to a model for a double neutron star system. Figure taken from Belczynski and Kalogera *Astrophysical Journal Letters* 550, L183 (2001). Reproduced by permission of the AAS

Problem 13.2: Determine the combination of the efficiency parameter and the structure parameter, $\alpha\lambda$ used for the common envelope phase from step (II) to (III), using the values from the microquasar example.

The common envelope phase happens on a dynamical timescale, so the secondary is almost unaffected. On the other hand, the primary now consists of a massive helium core with no surrounding envelope of hydrogen. Clearly, its further evolution will be dramatically altered by this change in its structure and composition. Such a star is called a helium star. It has sufficient mass to drive successive nuclear burning phases until an iron core develops. The helium star loses nearly $3\,M_\odot$ through stellar winds during this phase of its evolution, but when the core collapses there is no supernova. The star is massive enough to implode directly to a black hole. With no mass lost through the collapse event, the properties of the binary remain unchanged. The system maintains its configuration while the secondary continues to evolve along the main sequence. After another 5 Gyr, with the hydrogen exhausted in its core, the secondary ascends up the red giant branch and eventually fills its Roche lobe. The mass ratio is now very small ($q = 1.22/13.04 \sim 0.09$), and so the mass transfer is dynamically stable and proceeds on a thermal timescale. Since the mass transfer now proceeds from the lower-mass star to the higher-mass star, the orbital separation increases as does the orbital period. The mass transfer drives two jets from the accretion onto the black hole. The evolutionary phases are shown in Fig. 13.6.

Problem 13.3: Assume an initial period of 26.1 days for a binary containing a black hole with mass $m_1 = 13.04\,M_\odot$ and a red giant star with mass $m_2 = 1.22\,M_\odot$. If the binary is undergoing stable, conservative mass transfer from the red giant to the black hole with a mass transfer rate of $\dot{m}_2 = 1.0 \times 10^{-8}\,M_\odot$/year, what are the masses and orbital period after 10.5 Myr?

There are a few binary neutron star systems in the Milky Way. The formation of one is modeled from an initially eccentric binary containing two stars with ZAMS masses of $m_1 = 12.8\,M_\odot$ and $m_2 = 11.9\,M_\odot$ with a semimajor axis of $a = 830\,R_\odot$ and eccentricity of $e = 0.7$.

As the stars evolve through the main sequence, they lose mass through a stellar wind. When the more massive star evolves off the main sequence to become a giant, it circularizes the orbit as described in the example above, resulting in a semimajor axis of $a = 380\,R_\odot$. At the onset of Roche lobe overflow from the primary, the masses are $m_1 = 12.5\,M_\odot$ and $m_2 = 11.8\,M_\odot$. The mass ratio at this point is nearly 1 and so the mass transfer is dynamically stable, but nonconservative. The mass loss from the system tends to widen the system, but the mass transfer from the more massive to the less massive star tends to bring the stars together. The result is a loss of $4.7\,M_\odot$ from the system and a transfer of $4.8\,M_\odot$ to the secondary. The primary is now a helium core with a mass of $m_1 = 3.0\,M_\odot$ and the secondary has grown to $m_2 = 16.6\,M_\odot$. When the secondary evolves off the main sequence, the mass transfer is dynamically unstable due to the large mass ratio, and a common envelope develops. After the envelope is ejected, the system consists of two helium stars with masses $m_1 = 2.8\,M_\odot$ and $m_2 = 4.3\,M_\odot$, and an orbital separation of $5\,R_\odot$. These two stars evolve through core helium burning and eventually develop CO cores with convective envelopes. The expansion of the stars brings them into contact at roughly the same time, leading to a double common envelope. Both envelopes are ejected, leaving two CO cores, with masses of $m_1 = 1.7\,M_\odot$ and $m_2 = 2.4\,M_\odot$. The orbital separation is now $a = 0.5\,R_\odot$. These two cores subsequently explode in type Ic supernovae, which occur in stars that have lost their hydrogen envelopes. The remnant is a tight double neutron star binary. The evolutionary phases are shown in Fig. 13.7.

Problems

13.1. Assuming no mass loss from the stellar winds, use conservation of angular momentum to compute the semimajor axis of the tidally circularized orbit for step (I) to (II) in the microquasar example. How does it compare to the value of $3{,}501\,R_\odot$ obtained from the results of the stellar evolution code?

13.2. Determine the combination of the efficiency parameter and the structure parameter, $\alpha\lambda$ used for the common envelope phase from step (II) to (III), using the values from the microquasar example.

13.3. Assume an initial period of 26.1 days for a binary containing a black hole with mass $m_1 = 13.04\,M_\odot$ and a red giant star with mass $m_2 = 1.22\,M_\odot$. If the binary is undergoing stable, conservative mass transfer from the red giant to the black hole with a mass transfer rate of $\dot{m}_2 = 1.0 \times 10^{-8}\,M_\odot/\text{year}$, what are the masses and orbital period after 10.5 Myr?

Chapter 14
Star Cluster Dynamics

Many stars are born in clusters arising from the collapse and subsequent fragmentation of giant molecular clouds. During the birth process, radiation pressure from the ignition of bright massive stars and the pressure waves from the following supernovae can sweep away residual mass in the cloud. In some cases, the cluster will become unbound due to the loss of mass. The constituent stars of the cluster are then scattered and they continue to evolve in isolation. Some clusters are more massive and have greater central densities, so that they survive the initial mass loss. In these clusters, even the single stars may not evolve in isolation as they interact with the other stars in the cluster. The archetypical clusters of this type are globular clusters. In this chapter, we will cover some of the basics of modeling the dynamical evolution of these clusters.

14.1 Cluster Timescales

The evolution of star clusters depends on three basic timescales. The first timescale is the typical evolution time of the member stars of the cluster. This is comparable to the nuclear timescale. The next timescale is the *crossing time*, which measures the typical time required for a star to move through a characteristic radius of the cluster. Thus,

$$t_{\mathrm{cr}} = \frac{r}{v}, \tag{14.1}$$

where v is a typical velocity and r is a characteristic distance. Since different clusters do not have well-defined surfaces, the characteristic distance could refer to several length scales. Globular clusters are roughly spherical distributions of stars, with three basic radii that can be defined either observationally or theoretically. The innermost radius is the *core radius*, r_{c}. This radius is defined observationally as

M. Benacquista, *An Introduction to the Evolution of Single and Binary Stars*, Undergraduate Lecture Notes in Physics, DOI 10.1007/978-1-4419-9991-7_14, © Springer Science+Business Media New York 2013

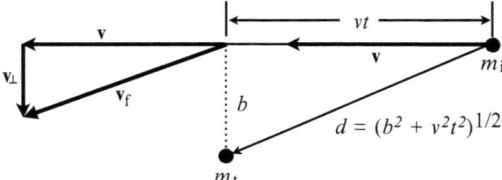

Fig. 14.1 The configuration for a weak encounter. The incident star m_i comes from the *right* with speed v and impact parameter b. After the encounter, the star leaves to the *left* with an additional velocity component $v_\perp$

the radius at which the surface brightness of the cluster drops to one-half of the central value. The theoretical definition of r_c is the radius at which the space density of stars drops to one-third of the central value. The *half-mass radius*, r_h, is defined theoretically as the radius of the sphere that contains half the mass of the system. Its observational counterpart is the *half-light radius* that contains half the light of the system. This is defined in terms of the integrated luminosity of the cluster. If the nature of the stellar population is dependent upon the radius, it is likely that the half-light radius and the half-mass radius will not coincide. The third length scale is the *tidal radius*, r_t, at which the host Galaxy's gravitational field dominates over the cluster's field. The typical velocity of a star will vary depending on its location in the cluster, and so the crossing time has different values for different locations in the cluster.

The third important timescale is the *relaxation time*, t_{rlx}, which measures the typical time required for the velocity of a star to change by an amount on the order of magnitude of itself. Another way of interpreting the relaxation time is that it is the time required for gravitational interactions with other stars to completely erase any trace of a star's initial orbit. The relaxation time can be computed by considering the time interval between interactions with other stars, combined with the size of the deflection expected to arise from such interactions.

A typical interaction in a cluster is a weak encounter where a star of mass m_i and speed v is incident on a stationary target star of mass m_t with an impact parameter b, as shown in Fig. 14.1. The time-dependent force on the incident star is

$$F(t) = \frac{Gm_t m_i}{d^2} = \frac{Gm_t m_i}{(b^2 + v^2 t^2)} \tag{14.2}$$

where $t = 0$ at the point of closest approach. We ignore the strong encounters, so we assume only a slight deflection. Therefore, after the encounter, a small velocity component, $v_\perp$, perpendicular to the initial velocity has been added to the velocity of the incident star. We compute $v_\perp$ using Newton's law:

$$m_i \frac{dv_\perp}{dt} = F_\perp = F \sin\theta = \frac{Gm_t m_i b}{(b^2 + v^2 t^2)^{3/2}}, \tag{14.3}$$

14.1 Cluster Timescales

Fig. 14.2 The cylindrical shell of width db, radius b, and length vt containing stars that will engage in a weak encounter with the incident star m_i

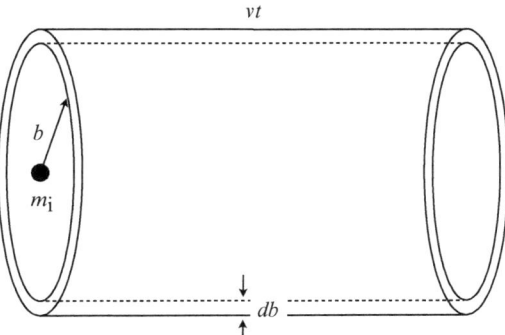

which integrates to

$$v_\perp = \int_{-\infty}^{\infty} \frac{Gm_t b}{(b^2 + v^2 t^2)^{3/2}} = \frac{2Gm_t}{bv}. \tag{14.4}$$

Each interaction will randomly kick the star, so the cumulative effect is like a random walk. This means that the kicks add in quadrature, so the net change in velocity after N weak encounters is the square root of the sum of the squares of the kicks:

$$\Delta v^2 = \sum_{i=1}^{N} (\Delta v_i)^2. \tag{14.5}$$

Accurately computing this sum requires knowing the mass and impact parameter for each target mass. If we assume some sort of averaged value of the mass, we can treat the problem as if there were identical target masses. In this case, then we can convert the sum to an integral over impact parameters. The fraction of encounters, dN, is related to the number density of stars, n, and the volume of a cylindrical shell of length vt, radius b, and thickness db through

$$dN = ndV = nvt 2\pi b db, \tag{14.6}$$

as shown in Fig. 14.2. Therefore, the average change in the velocity through weak encounters is found from

$$\langle \Delta v^2 \rangle = \int \Delta v^2 dN = \int_{b_{\min}}^{b_{\max}} \left(\frac{2Gm_t}{bv}\right)^2 nvt 2\pi b \, db = \frac{8\pi G^2 m_t^2 nt}{v} \ln\left(\frac{b_{\max}}{b_{\min}}\right), \tag{14.7}$$

where the limits of integration give the minimum impact parameter where the weak limit is valid ($b_{\min}$) and the maximum impact parameter ($b_{\max}$) that is usually the size of the cluster. Remembering that the relaxation time is the typical time required for the velocity of a star to change by an order of magnitude of itself, we can set $\langle \Delta v^2 \rangle = v^2$ to obtain

$$t_{\text{rlx}} = \frac{v^3}{8\pi G^2 m_t^2 n \ln(b_{\max}/b_{\min})}. \tag{14.8}$$

The logarithm in the denominator is known as the *Coulomb logarithm* and can also be described in terms of N, the total number of stars in the cluster, as $\ln \gamma N$, where γ is an empirically determined number that depends on the mass distribution of the system. Typical values of γ range from 0.02 to 0.4.

The numerical values of these timescales depend upon the typical masses, velocities, and number densities of the constituent stars in the cluster. Assuming a standard distribution of initial masses of stars in the cluster, the current mass distribution will depend upon the age of the cluster and the details of stellar evolution. The typical velocity of a star in the cluster can be obtained from a version of the virial theorem applied to stars in a self-gravitating system. The treatment of the virial theorem in Chap. 4 assumed that the gas in a star was in hydrostatic equilibrium. Here we will find a less restrictive requirement to relate the average kinetic energy of a star in a cluster to the gravitational potential of the cluster.

To obtain the virial theorem for clusters, we start with the moment of inertia evaluated about a point for a cluster of point masses:

$$I = \sum_i m_i r_i^2 = \sum_i m_i (\mathbf{r}_i \cdot \mathbf{r}_i) \tag{14.9}$$

and take two derivatives with respect to time to obtain

$$\frac{d^2 I}{dt^2} = 2\sum_i m_i \left(v_i^2 + \mathbf{r}_i \cdot \mathbf{a}_i\right). \tag{14.10}$$

Since the dynamics of the cluster is completely determined by gravitation, we can relate the acceleration to the gravitational potential via

$$m_i \mathbf{a}_i = -\nabla_i \Omega(\mathbf{r}_i) \tag{14.11}$$

where ∇_i indicates the gradient taken with respect to the coordinates of m_i and the gravitational potential at $\mathbf{r}_i$ is

$$\Omega(\mathbf{r}_i) = \sum_{j \neq i} \frac{G m_j m_i}{|\mathbf{r}_i - \mathbf{r}_j|}. \tag{14.12}$$

Using this expression for the acceleration, Eq. (14.10) reads

$$\frac{d^2 I}{dt^2} = 2\sum_i m_i v_i^2 - 2\sum_i \mathbf{r}_i \cdot \nabla_i \left(\sum_{j \neq i} \frac{G m_j m_i}{|\mathbf{r}_i - \mathbf{r}_j|}\right). \tag{14.13}$$

14.1 Cluster Timescales

Because the last sum on the right-hand side is over both i and j, we can exchange the indices to find

$$2\sum_i \mathbf{r}_i \cdot \mathbf{V}_i \left(\sum_{j\neq i} \frac{Gm_jm_i}{|\mathbf{r}_i - \mathbf{r}_j|} \right) = 2\sum_{i\neq j} \left(\mathbf{r}_i \mathbf{V}_i \frac{Gm_jm_i}{|\mathbf{r}_i - \mathbf{r}_j|} + \mathbf{r}_j \mathbf{V}_j \frac{Gm_jm_i}{|\mathbf{r}_j - \mathbf{r}_i|} \right)$$

$$= -2\sum_{i\neq j} \frac{Gm_jm_i}{|\mathbf{r}_i - \mathbf{r}_j|} = -2\Omega, \quad (14.14)$$

where Ω is the total gravitational potential energy of the cluster. The total kinetic energy of the cluster is

$$K = \frac{1}{2}\sum_i m_i v_i^2, \quad (14.15)$$

so the second derivative of the moment of inertia of the cluster is then

$$\frac{d^2I}{dt^2} = 4K + 2\Omega. \quad (14.16)$$

We can see that if the moment of inertia of the cluster varies linearly with time, then the kinetic and potential energies of the cluster obey a virial theorem:

$$K = -\frac{1}{2}\Omega. \quad (14.17)$$

Problem 14.1 Evaluate the gradients in Eq. (14.14) to show that

$$2\sum_i \mathbf{r}_i \cdot \nabla_i \Omega(\mathbf{r}_i) = -2\Omega.$$

As the cluster evolves, it slowly loses mass as individual stars randomly acquire escape velocity through interactions. This is analogous to the evaporation of a gas. The speeds of the stars can be modeled using a Maxwell–Boltzmann distribution, and the relaxation process tends to maintain the Maxwell–Boltzmann distribution. The average escape velocity for stars in a cluster is twice the average speed, so $\langle v_e^2 \rangle = 4\langle v^2 \rangle$. From the Maxwell–Boltzmann distribution, the fraction of stars having a speed $v > 2\langle v^2 \rangle$ is $\delta = 0.00738$, and so we can assume that this fraction of stars escapes the cluster during each relaxation time. Therefore, we define the *evaporation time* as

$$t_{\text{evap}} = \frac{t_{\text{rlx}}}{\delta} = 136 t_{\text{rlx}}. \quad (14.18)$$

The evaporation time gives a timescale for the life of the cluster before it disappears through this process. Note that this definition of the evaporation time does not include the effect of any external potential, which would tend to reduce the evaporation time.

From the virial theorem, we can calculate the average velocity in a cluster and get estimates for the crossing time and the relaxation time. If we assume an average mass $\langle m \rangle$ and root-mean-square velocity $\langle v^2 \rangle$, then Eq. (14.17) can be approximated by

$$\frac{1}{2}N\langle m\rangle\langle v^2\rangle = \frac{1}{2}\frac{GN^2\langle m\rangle^2}{r}, \qquad (14.19)$$

where N is the total number of stars in the cluster and r is the typical radius of the cluster. Thus, the typical velocity in Eq. (14.1) is

$$v = \sqrt{\langle v^2 \rangle} \simeq \sqrt{\frac{GN\langle m \rangle}{r}}. \qquad (14.20)$$

Approximating the density as $\rho = N\langle m \rangle/r^3$, we find

$$t_{\rm cr} \simeq \frac{1}{\sqrt{G\rho}}, \qquad (14.21)$$

which is of the same form as the dynamical timescale from Eq. (4.64). Note that this approximation gives a local definition of the crossing time. We see that the crossing time is shorter in the core of a cluster where the density is greater. Using the virial theorem in conjunction with the relaxation time, we find that

$$t_{\rm rlx} \simeq \frac{N}{8\pi \ln \gamma N} \frac{1}{\sqrt{G\rho}} \sim \frac{0.1N}{\ln \gamma N} t_{\rm cr}. \qquad (14.22)$$

Problem 14.2 Using Eq. (14.20) and an estimate of the density, derive Eq. (14.22).

Open clusters, with $N \sim 10^2$, have $t_{\rm rlx} < 10~t_{\rm cr}$. Therefore a star in an open cluster will have its motion perturbed from what you would expect for a point particle in a smooth potential after a few crossings. A typical open cluster has a mass of a few hundred solar masses and a central density of about $100\,M_\odot/{\rm pc}^3$. Therefore, the crossing time is around $t_{\rm cr} = 2 \times 10^6$ year, and the relaxation time is $t_{\rm rlx} = 2 \times 10^7$ year. Consequently, the evaporation time is $136 t_{\rm rlx} = 2.7 \times 10^9$ year. Open clusters tend to dissolve in about three billion years, releasing their stars to the disk or field population. This timescale is less than the main sequence lifetime of many of the low-mass stars and substantially less than the age of the Galaxy. Thus, open clusters tend to be young and the likelihood of strong stellar interactions within them is quite low.

Globular clusters, with $N \sim 10^6$, tend to have $t_{\rm rlx} \sim 10^4~t_{\rm cr}$. This means that stars in globular clusters make many crossings with little significant change in their orbits due to weak interactions. A typical globular cluster has a total mass of $10^5\,M_\odot$ and a central density of nearly $10^4\,M_\odot/{\rm pc}^3$. The crossing time is around

14.2 Globular Cluster Structure

Table 14.1 Timescales and properties for open and globular clusters

Type	M_{tot} ($M_\odot$)	ρ_0 ($M_\odot/pc^3$)	N	t_{cr} (year)	t_{rlx} (year)	t_{evap} (year)
Open	250	100	100	2×10^6	2×10^7	2.7×10^9
Globular	6×10^5	8,000	10^6	1.5×10^5	1.5×10^9	2×10^{11}

$t_{cr} = 1.5 \times 10^5$ year, and so the relaxation time is $t_{rlx} = 1.5 \times 10^9$ year. From this, we find the evaporation time to be $t_{evap} = 2 \times 10^{11}$ year, which is greater than the age of the universe and greater than the lifetime of most stars. However, tidal interactions between globular clusters and their host galaxies can significantly enhance evaporation rates. Typical properties and timescales for open and globular clusters are shown in Table 14.1.

14.2 Globular Cluster Structure

The timescales associated with open clusters imply that such clusters have little or no influence on the formation of binaries and hence the evolution of stars. Therefore, stars in open clusters evolve similarly to isolated field stars. On the other hand, globular clusters have much higher densities and longer lives, and so they can influence the evolution of their constituent stars. Because of the significant difference between the crossing time and the relaxation time, it is possible to break the evolution of clusters into two separate regimes. On timescales shorter than the relaxation time, the cluster potential can be treated as a smooth, static background potential in which the cluster stars behave as a collisionless gas. On longer timescales, the background potential also evolves as weak interactions between the stars become important. In this case the stars behave as a collisional gas. In this section, we will concentrate on the short timescales and the collisionless gas model in order to develop models of the cluster potential.

The challenge of determining the structure of a globular cluster lies in the fact that the potential is determined from the distribution of stars in the cluster, but the distribution of stars in the cluster is determined by the potential. Therefore, we need to find a self-consistent way of relating the dynamics of cluster stars with the potential. One way of doing this is to model the cluster as a collisionless gas where each star is treated as a gas particle. The properties of the cluster stars are described using a *distribution function* $f(\mathbf{r}, \mathbf{v}, m)$ that describes the probability of finding a star of mass m at a particular location in a six-dimensional position-velocity phase space. When normalized to the total number of stars in the cluster (N), the distribution function can also be interpreted as the density of stars of a given mass with a given velocity and position:

$$\rho(\mathbf{r}, \mathbf{v}, m) d^3v d^3r dm = N f(\mathbf{r}, \mathbf{v}, m) d^3v d^3r dm. \qquad (14.23)$$

If the dynamics of the cluster are governed by simple Newtonian gravity, we can consider f to be a conserved quantity in phase space, and so the continuity equation for f reads

$$\frac{df}{dt} = \frac{\partial f}{\partial t} + \mathbf{v} \cdot \nabla_{\mathbf{r}} f + \mathbf{a} \cdot \nabla_{\mathbf{v}} f = 0, \tag{14.24}$$

where

$$\mathbf{v} \cdot \nabla_{\mathbf{r}} f = \sum_{i=1}^{3} v_i \frac{\partial f}{\partial x_i}, \tag{14.25}$$

$$\mathbf{a} \cdot \nabla_{\mathbf{v}} f = \sum_{i=1}^{3} a_i \frac{\partial f}{\partial v_i}. \tag{14.26}$$

If we now define the gravitational potential Φ, using

$$\nabla^2 \Phi = 4\pi G \rho(\mathbf{r}), \tag{14.27}$$

where

$$\rho = \int f d^3 v dm, \tag{14.28}$$

then the acceleration is

$$\mathbf{a} = -\nabla_{\mathbf{r}} \Phi. \tag{14.29}$$

Thus, the evolution of f is governed by the *collisionless Boltzmann equation*:

$$\frac{\partial f}{\partial t} + \mathbf{v} \cdot \nabla_{\mathbf{r}} f - \nabla_{\mathbf{r}} \Phi \cdot \nabla_{\mathbf{v}} f = 0. \tag{14.30}$$

Globular clusters are nearly spherical, so if we assume spherical symmetry as we did for stellar evolution and we require a stationary solution, then we can substantially simplify the problem of solving Eq. (14.30) for Φ. In a spherically symmetric, stationary system, the distribution function can only depend on one variable. Since energy is conserved throughout the orbit, then $E(\mathbf{r}, \mathbf{v})$ obeys

$$\frac{dE}{dt} = 0 = \frac{\partial E}{\partial t} + \mathbf{v} \cdot \nabla_{\mathbf{r}} E - \nabla_{\mathbf{r}} \Phi \cdot \nabla_{\mathbf{v}} E, \tag{14.31}$$

and so E is, itself, a solution to Eq. (14.30). For the spherically symmetric case, this implies that any distribution function of the form $f(E)$ is also a stationary solution to the collisionless Boltzmann equation. It follows that one approach to finding a self-consistent solution to Eq. (14.30) is to propose a functional form for $f(E)$ and then determine the form of Φ by obtaining ρ and solving the Poisson equation. Before doing this, we will introduce two functions of the potential and the energy. This is done to obtain a solution in which the argument of the distribution function

14.2 Globular Cluster Structure

behaves kind of like a radius (i.e., it is positive everywhere). The relative potential is defined as

$$\Psi = -\Phi + \Phi_0, \tag{14.32}$$

where Φ_0 is chosen so that the relative energy

$$\mathcal{E} = -E + \Phi_0 \tag{14.33}$$

is always greater than 0. The relative energy can also be written in terms of Ψ and v as

$$\mathcal{E} = \Psi - \frac{1}{2}v^2. \tag{14.34}$$

Since $\mathcal{E}$ is a simple linear function of E, we write the distribution function as

$$f(\mathcal{E}) = F\mathcal{E}^{n-3/2}. \tag{14.35}$$

The peculiar choice of the exponent is to simplify the equations later on. Computing the density from this distribution function gives:

$$\rho(\mathbf{r}) = \int_0^\infty \int f(\mathcal{E}) \mathrm{d}^3 v \, \mathrm{d}m = C_n \Psi^n. \tag{14.36}$$

Taking this density, we can write the Poisson equation in spherical coordinates with spherical symmetry to find

$$\frac{1}{r^2}\frac{\mathrm{d}}{\mathrm{d}r}\left(r^2 \frac{\mathrm{d}\Psi}{\mathrm{d}r}\right) = -4\pi G C_n \Psi^n, \tag{14.37}$$

which is dangerously close to the Lane–Emden equation! Following the procedure in Chap. 8, we first rewrite everything in a dimensionless form. Define $\Psi_0 = \Psi(0)$, and introduce the dimensionless potential:

$$\phi = \frac{\Psi}{\Psi_0}. \tag{14.38}$$

Next, noting that $4\pi G C_n \Psi_0^{n-1}$ must have dimensions of length, we introduce the characteristic length

$$d = \frac{1}{\sqrt{4\pi G C_n \Psi_0^{n-1}}} \tag{14.39}$$

and the dimensionless variable $s = r/d$. With these substitutions, we recover the Lane–Emden equation:

$$\frac{1}{s^2}\frac{\mathrm{d}}{\mathrm{d}s}\left(s^2 \frac{\mathrm{d}\phi}{\mathrm{d}s}\right) = -\phi^n, \tag{14.40}$$

with the boundary conditions

$$\phi(0) = 1, \tag{14.41}$$

$$\left.\frac{d\phi}{ds}\right|_0 = 0. \tag{14.42}$$

In this case, the variable s corresponds to radius and the variable ϕ is related to the density through $\rho = C_n \Psi_0^m \phi^n$. We have seen from the solutions to the Lane–Emden equation for polytropes that $1 \leq n < 5$ produce objects with finite mass and finite radius. Solutions with $n \geq 5$ yield objects with finite mass, but infinite radius. Since globular clusters do not have well-defined surfaces, solutions with $n \geq 5$ will describe these objects. The only analytic solution to the Lane–Emden equation is for $n = 5$, and so this solution is frequently used to describe the mass distribution and potential of globular clusters on timescales much shorter than the relaxation time. This solution is called the *Plummer model*.

The $n = 5$ solution to Eq. (14.40) is

$$\phi(s) = \left(1 + \frac{s^2}{3}\right)^{-1/2}. \tag{14.43}$$

If we define the *Plummer radius* to be $a = \sqrt{3}d$, then for a cluster of mass M, the stellar density is

$$\rho(r) = \frac{3Ma^2}{4\pi\sqrt{(r^2 + a^2)^5}} \tag{14.44}$$

and the associated potential is

$$\Phi(r) = -\frac{GM}{\sqrt{r^2 + a^2}}. \tag{14.45}$$

The Plummer distribution function is $f(\mathcal{E}) = \mathcal{E}^{7/2}$. The Plummer model is one of many models that can be used to describe the orbits and mass distribution of stars in globular clusters over timescales that are short compared to the relaxation time, so that the cluster can be treated as a stationary system. A synthetic globular cluster using a Plummer model is compared with the globular cluster M80 in Fig. 14.3. Over longer timescales, it will be necessary to include the evolution of the cluster.

Problem 14.3: Using the density of the Plummer model, compute the core radius r_c using the theoretical definition.

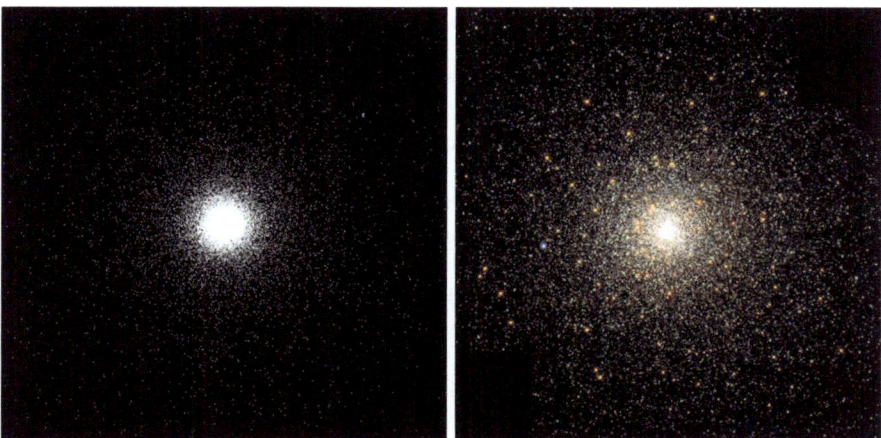

Fig. 14.3 Comparison of the Plummer model (*left*) with the globular cluster M80 (*right*). Photo courtesy of NASA

Problem 14.4: Use the density of the Plummer model.

(a) Compute the surface density of stars using

$$\Sigma(R)R\,\mathrm{d}R\,\mathrm{d}\phi = \int_{-\infty}^{+\infty} \rho(\sqrt{R^2+z^2})R\,\mathrm{d}R\,\mathrm{d}\phi\,\mathrm{d}z,$$

where the integration is over z. R and ϕ are cylindrical coordinates.

(b) Compute the core radius r_c using the observational definition.

14.3 Globular Cluster Evolution

On longer timescales that are comparable to the relaxation time or the stellar lifetime, then the cluster structure evolves and we can no longer assume a stationary distribution function or background potential. In effect, the cluster becomes a *collisional* gas. This requires a modification of the Boltzmann equation to include a term governing the effect of stellar interactions on the distribution function. The *collisional Boltzmann equation* takes the form

$$\frac{\partial f}{\partial t} + \mathbf{v}\cdot\nabla_{\mathbf{r}}f - \nabla_{\mathbf{r}}\Phi\cdot\nabla_{\mathbf{v}}f = \Gamma(f). \tag{14.46}$$

It is not possible to evaluate Γ analytically, and so it is approximated by a variety of numerical methods.

The most direct method of solution is to simply numerically compute the trajectories of every star in the cluster. Since the interactions are entirely determined by gravitational interactions, this approach is conceptually the simplest, but it entails a large number of computations for each time step and is hindered by the amount of computing time required for realistic clusters. Other approaches involve approximating Γ in the weak-scattering limit or by a Monte Carlo selection of weak encounters of timescales shorter than the relaxation time. In any case, these approaches are beyond the scope of this book.

The primary interest here is how membership in a cluster alters the evolution of binary and single stars. Therefore, we will look at specific features of cluster evolution that result in an increase in the number density of stars and therefore cause the approximation of isolation that we have used for both single and binary stars to break down.

At the birth of a cluster, the more massive stars will ignite first. The resulting stellar winds and radiation pressure will expel some of the gas in the cluster. As a result of the relatively rapid loss of mass, the potential changes quickly and the stars in the cluster are thrown out of virial equilibrium. This means that the positions and velocities are uncorrelated and independent of the masses of the stars. This initial process is called *violent relaxation*.

After violent relaxation, the cluster begins to return to virial equilibrium as encounters between stars distribute the energy equally among all stars. This equipartition of energy causes stars with large kinetic energies to transfer their energy to stars with kinetic energies that are below the average kinetic energy. In general, this means that more massive stars tend to end up with lower velocities than low-mass stars. Thus, the more massive stars tend to sink toward the center of the cluster while low-mass stars migrate to the outer halo of the cluster. This process is known as *mass segregation*. During mass segregation, the more massive stars are losing kinetic energy to the less massive stars through weak encounters. Therefore the timescale for equipartition is related to t_{rlx}. Numerical simulations show that for massive stars of m_i, the equipartition timescale is $t_{\text{eq}} \propto t_{\text{rlx}} (\langle m \rangle / m_i)$, where $\langle m \rangle$ is the average mass of a star in the cluster.

Mass segregation can become unstable so that equipartition is never reached. In this case, the most massive stars sink to the core of the cluster and effectively decouple from the rest of the stars. This is known as the *Spitzer instability*. The important features of the Spitzer instability can be seen in a simple two-component model of a cluster. Consider a cluster consisting of two populations of stars. There are N_1 stars with mass m_1 and N_2 stars with mass m_2, with $m_2 > m_1$. The total mass of each population is given by $M_i = N_i m_i$ where $i = 1, 2$. If we assume that the two populations have reached equipartition, then

$$m_1 \langle v_1^2 \rangle = m_2 \langle v_2^2 \rangle. \tag{14.47}$$

14.3 Globular Cluster Evolution

Since each population would also be in virial equilibrium, they also obey

$$\langle v_1^2 \rangle = \frac{\alpha G M_1}{r_{h1}} + \frac{G}{M_1} \int_0^\infty \frac{\rho_1 M_2(r)}{r} 4\pi r^2 dr, \quad (14.48)$$

$$\langle v_2^2 \rangle = \frac{\alpha G M_2}{r_{h2}} + \frac{G}{M_2} \int_0^\infty \frac{\rho_2 M_1(r)}{r} 4\pi r^2 dr, \quad (14.49)$$

where ρ_i is the local density of stars of mass m_i, r_{hi} is the half-mass radius of population i, and $M_i(r)$ is the total mass of population i contained within radius r. The first term on the right-hand side represents the self-gravity of the population and α is a parameter that describes the density distribution throughout the cluster. For Plummer-type polytrope models, $\alpha = 0.38$. The second term on the right-hand side represents the gravitational energy of one population due to the other.

We assume that the more massive stars make up a small fraction of the total mass of the cluster, so $M_2 \ll M_1$. At the same time, we assume that the more massive stars are substantially more massive than the less massive stars, so $m_2 \gg m_1$. Furthermore, we expect the more massive stars to have segregated and become centrally concentrated compared to the distribution of low-mass stars. Therefore, we assume that

$$M_1(r) \simeq \frac{4}{3}\pi r^3 \rho_{c1}, \quad (14.50)$$

where ρ_{c1} is the central density of stars of mass m_1. With these assumptions, we can neglect the second term in Eq. (14.48), and so

$$\langle v_1^2 \rangle = \frac{\alpha G M_1}{r_{h1}}, \quad (14.51)$$

$$\langle v_2^2 \rangle = \frac{\alpha G M_2}{r_{h2}} + G\rho_{c1} \frac{4}{3}\pi r_{s2}^2, \quad (14.52)$$

where r_{s2}^2 represents some sort of integrated average value of r^2 for stars of mass m_2. Now, we define the mean density of stars of each type within their half-mass radius to be

$$\rho_{mi} = \frac{1.5 M_i}{4\pi r_{hi}^3}. \quad (14.53)$$

Using this equation to express r_{h2}/r_{h1} in terms of ρ_{m2}/ρ_{m1} and substituting the values of $\langle v_1^2 \rangle$ and $\langle v_2^2 \rangle$ from Eqs. (14.51) and (14.52) into the equipartition condition (Eq. (14.47)), we obtain

$$\frac{M_2}{M_1}\left(\frac{m_2}{m_1}\right)^{3/2} = \frac{(\rho_{m1}/\rho_{m2})^{1/2}}{(1+\beta(\rho_{m1}/\rho_{m2}))^{3/2}}, \quad (14.54)$$

where

$$\beta = \frac{\rho_{c1}}{\rho_{m1}} \frac{1}{2\alpha} \left(\frac{r_{s2}}{r_{h2}}\right)^2. \tag{14.55}$$

The left-hand side of Eq. (14.54) has a maximum value of $\gamma = \sqrt{4/27\beta}$ at $\rho_{m1}/\rho_{m2} = (2\beta)^{-1}$. Therefore, if

$$\frac{M_2}{M_1} > \gamma \left(\frac{m_2}{m_1}\right)^{3/2}, \tag{14.56}$$

then equipartition will never be reached and the stars in population two will decouple from the cluster. For typical cluster profiles modeled on polytropes with $3 < n < 5$, α varies over a small range of $0.38 < \alpha < 0.42$. For these same polytropes, the ratio of central density to the half-mass mean density ranges between 2.5 and 4.5. Therefore,

$$\gamma \simeq \frac{35}{8} \left(\frac{r_{s2}}{r_{h2}}\right)^2. \tag{14.57}$$

We would expect the ratio of radii to be of order 1, and for a Maxwell–Boltzmann distribution in a parabolic potential well, it is 0.9. Thus, $5 < \gamma < 6$, for reasonable values. In Spitzer unstable systems, massive stars sink to the center of the cluster and interact only with themselves.

Even if equipartition is reached, the evolution of the cluster slowly leads to a concentration of stars in the core of the cluster. This is a direct result of the fact that stellar clusters can be thought of as self-gravitating gas clouds where the stars play the role of gas particles. In this analogy, one can talk about a *dynamical temperature*, T, of the cluster, where

$$K = \frac{1}{2} \sum_i m_i v_i^2 = \frac{3}{2} NkT. \tag{14.58}$$

From the virial theorem, we see that the total energy of the cluster is

$$E = K + \Omega = -K = -\frac{3}{2} NkT. \tag{14.59}$$

As with stars in hydrostatic equilibrium, this implies a negative heat capacity, C:

$$C = \frac{dE}{dT} = -\frac{3}{2} Nk. \tag{14.60}$$

Therefore, as the cluster loses energy, it heats up.

The introduction of the collisional term in Eq. (14.46) is similar to adding a heat conduction term to a gas. In the core of the cluster, the stars are moving faster than the average speed for the cluster, and therefore the core can be considered to be "hotter" than the outer regions of the cluster. The collisional term allows this heat energy to be conducted to the outer regions of the cluster. Because of the negative heat capacity, this causes the inner regions to contract and heat up while the outer regions expand and cool. The cluster develops an unstable "core-halo" structure and the core continues to collapse while the halo expands and evaporates. Once the density in the core exceeds a limit set by $\rho_c = 709\rho_h$, then an instability known as the Antonov instability comes into play and rapidly drives the core to a singular solution in a process known as the *gravothermal catastrophe*. Numerical simulations have shown that this core collapse takes place over a timescale of a few tens of relaxation times. This process is similar to the collapse of a gas sphere to form a star. In the case of a star, the collapse is halted by a source of energy in the form of nuclear fusion. For a cluster, the source of energy is the gravitational binding energy released when binary stars are driven closer together through encounters in the core. In the next chapter, we will discuss both the dynamical formation and the disruption of binaries within stellar clusters.

Problems

14.1. Evaluate the gradients in Eq. (14.14) to show that

$$2\sum_i \mathbf{r}_i \cdot \nabla_i \Omega(\mathbf{r}_i) = -2\Omega.$$

14.2. Using Eq. (14.20), and an estimate of the density, derive Eq. (14.22).

14.3. Using the density of the Plummer model, compute the core radius r_c using the theoretical definition.

14.4. Use the density of the Plummer model.

(a) Compute the surface density of stars using

$$\Sigma(R)R\,dR\,d\phi = \int_{-\infty}^{+\infty} \rho(\sqrt{R^2+z^2})R\,dR\,d\phi\,dz,$$

where the integration is over z. R and ϕ are cylindrical coordinates.
(b) Compute the core radius r_c using the observational definition.

Chapter 15
Dynamical Evolution of Binaries

Stellar clusters are expected to be born with a fraction of their stars in binary systems. The primordial binary fraction is somewhat uncertain because the population of binaries is altered throughout the evolution of the cluster. Primordial binaries can be disrupted in the core of the cluster in order to provide the energy needed to halt core collapse. In addition, single stars can combine through strong interactions in order to form new binaries. In this final chapter, we will look at the ways in which dynamical interactions within clusters can alter the population of binary stellar systems within the cluster.

15.1 Dynamical Formation

Strong dynamical interactions between stars in a cluster can dissipate the energy of two stars so that they become gravitationally bound, forming a binary. Two of the most common mechanisms are tidal dissipation through two-body interactions and three-body interactions. Two-body interactions are much more common than three-body interactions; however, they require very close encounters in order to dissipate enough energy through tides to bind the system. Starting with two stars that are unbound, but still members of the cluster, we can determine the amount of energy that must be dissipated in tidal heating in order to form a binary. Although the system evolves from inside the cluster potential, we consider the two stars to be isolated with a relative velocity of v_∞ prior to the encounter. We can take $v_\infty = 10\,\text{km/s}$, which is a typical dispersion velocity of a star in a cluster. Therefore, in a coordinate system which is centered on one star, the total initial energy of the system is

$$E_i = \frac{1}{2}\mu v_\infty^2, \tag{15.1}$$

where μ is the reduced mass. This initial energy is the minimum amount of energy that must be dissipated in order for the encounter to produce a bound binary system. The closest point of the encounter, $r_{\min}$, will also be the periastron of the newly formed binary. At periastron, tidal forces will cause the stars to deviate from spherical symmetry. The *oblateness* in star i due to star j is

$$\varepsilon_{ij} = \frac{m_j}{m_i}\left(\frac{R_i}{r_{\min}}\right)^3, \tag{15.2}$$

where R_i is the undisturbed radius of star i. If we assume that the gravitational energy of this distortion is removed from the kinetic energy of the stars and periastron and then dissipated within each star, then the tidal interaction removes

$$\Delta E_t = \frac{Gm_1^2\varepsilon_{12}^2}{R_1} + \frac{Gm_2^2\varepsilon_{21}^2}{R_2} \tag{15.3}$$

from the system. Therefore a bound system will result when

$$\Delta E_t > \frac{1}{2}\mu v_\infty^2. \tag{15.4}$$

This places an upper bound on $r_{\min}$ in order for tidal dissipation to produce a binary of

$$r_{\min} < \left[\frac{2G}{\mu v_\infty^2}\left(m_2^2 R_1^5 + m_1^2 R_2^5\right)\right]^{1/6}. \tag{15.5}$$

We can relate this upper bound on $r_{\min}$ to an impact parameter, b. First we note that conservation of angular momentum requires $v_p r_{\min} = v_\infty b$, where v_p is the velocity at $r_{\min}$. The velocity at $r_{\min}$ can be found through conservation of energy, giving

$$v_p^2 = v_\infty^2\left(1 + \frac{2GM}{r_{\min}}\right) \simeq \frac{2GMv_\infty^2}{r_{\min}}, \tag{15.6}$$

where $M = m_1 + m_2$. Therefore,

$$b \simeq \sqrt{\frac{2GMr_{\min}}{v_\infty^2}} \tag{15.7}$$

and the cross section for an encounter resulting in tidal capture is

$$\sigma = \pi b^2 \simeq 2\pi\frac{GMr_{\min}}{v_\infty^2}. \tag{15.8}$$

From here, we can compute the rate of tidal capture encounters in terms of the density and number of stars in the center of a cluster:

$$\dot{N} = N\sigma v_\infty n = 2\pi Nn\frac{G(m_1+m_2)}{v_\infty}\left[\frac{2G}{\mu v_\infty^2}\left(m_2^2 R_1^5 + m_1^2 R_2^5\right)\right]^{1/6}. \tag{15.9}$$

15.1 Dynamical Formation

Typically, the binaries that are formed through tidal capture encounters have very high eccentricity and are marginally bound. As we shall see later, such binaries are frequently disrupted in later encounters.

Three-body (and higher order) encounters can produce binaries as the bodies exchange kinetic energy during close interactions. A qualitative description of the process can be obtained by considering the few-body interaction in isolation. Restricting ourselves to a three-body encounter for simplicity, the three bodies are marginally unbound at the start of the encounter, so we take the total energy to be ~ 0. During the encounter, the kinetic energy of the three bodies will try to equalize. In this case, the lowest mass object is most likely to gain the most velocity through the encounter. If this velocity exceeds the escape velocity of the three-body system, it can carry away enough energy so that the total energy of the remaining two stars is negative and they become bound. This qualitative argument is based on a statistical description of the outcome of many three-body interactions. In practice, many three-body encounters simply leave three unbound stars. In the cases where a binary is formed, it is not necessary that the lowest mass star is the one ejected, but it is the most common outcome.

The production rate of binaries formed through three-body encounters can be described in terms of the number density of single stars and a rate function Q. Since the density of stars within a cluster has a radial dependence, we write the number density as $n(\mathbf{R})$, where $\mathbf{R}$ is the radial distance from the center of the cluster. The rate function depends on the dynamics of the three-body interaction and can be expressed in terms of x, the relative binding energy of the resulting binary. This is written in such a way so that $x > 0$, so

$$x = -\frac{1}{2}\mu v^2 + \frac{GM\mu}{r}, \tag{15.10}$$

where $M = m_1 + m_2$, $\mu = m_1 m_2 / M$, and $v = \dot{r}$. Because this is a three-body process, the rate of formation of binaries with binding energies in the range x to $x + dx$ depends on the cube of the number density, and it is

$$\dot{n}(\mathbf{R}, x) dx = n^3(\mathbf{R}) Q(x) dx. \tag{15.11}$$

Evaluation of $Q(x)$ requires a knowledge of the distribution functions for stars in the cluster, and so a detailed computation of Q is beyond the scope of this book. For many reasonable cluster models, $Q(x) \propto x^{-7/2}$. This implies that most binaries that are formed through three-body encounters have small binding energies and so they are loosely bound. Since the formation rate also depends on n^3, most binaries are formed in the central regions of clusters where the number densities are the greatest. The rate function is also independent of the eccentricity and so the eccentricities of these binaries follow a distribution that has been determined from numerical simulations to be

$$f(e) = 2e. \tag{15.12}$$

This is also known as a *thermal distribution*.

> **Problem 15.1:** Calculate the average and median values of the eccentricity for a population of binaries with a thermal distribution of eccentricities.

Binaries formed through dynamical interactions are thought to be quite rare due to the close encounters required for tidal dissipation and the reduced likelihood of three bodies interacting at once. Furthermore, both formation scenarios tend to produce weakly bound, eccentric systems. Such systems are likely to interact later with other stars in the cluster and these interactions tend to disrupt weakly bound systems. Thus, we expect most binaries in a cluster to begin their lives as primordial binaries. The interaction of binary systems within a cluster is discussed in the next section.

15.2 Binary Interactions

The population of stars within a cluster will contain single stars and binary stars. Although some of the binary stars in a cluster will be formed through the dynamical processes previously discussed, most of the binary systems are expected to be primordial. In the crowded environment found in the center of a cluster, these systems are highly likely to interact with one another. These interactions will alter the properties of the binaries throughout the evolution of the cluster, resulting in binary systems that cannot have formed in isolation. The details of these interactions are not amenable to a simple quantitative description, and so the expected outcomes are usually determined through numerical solution of the equations of motion for few-body systems. A brief introduction to numerical methods will be discussed in the next section, while here we will use more qualitative, statistical arguments to describe likely outcomes of these interactions.

The most common form of interaction involving a binary system will be a binary–single encounter. In these encounters, there are two important variables, the relative binding energy of the binary, x, and the relative kinetic energy of the single star, $m_s v^2/2$. In a close encounter, the components of the binary will exchange energy with the single star, tending to equally distribute the kinetic energy between all three stars. The kinetic energy of the components of the binary is directly related to the relative binding energy through the virial theorem. Therefore, if the single star comes in with a kinetic energy greater than the binding energy then it will increase the kinetic energy of the components of the binary and the net result will be a widening of the binary and a net reduction of the relative binding energy. (Remember that the binding energy is actually negative, so the addition of energy to the systems results in a reduction of the relative binding energy.) Conversely, if the single star comes in with a kinetic energy less than the binding energy, then it will gain energy from the binary and leave the binary in a tighter orbit with a greater relative binding energy. At a crude level, where we assume that the mass of every star in the system is equal to the average mass, the average incoming kinetic

15.2 Binary Interactions

energy of the single star will be related to the dynamical temperature defined in Eq. (14.58), so $m_s v^2 \simeq 3kT$. Therefore, binary systems with $x < kT$, will tend to be driven to ever smaller values of x through repeated encounters. Such systems are said to be "soft" because they are likely to be disrupted over time. On the other hand, binary systems with $x > kT$ will have their relative binding energy increased through repeated encounters, becoming more tightly bound. These systems are called "hard" binaries. This positive feedback results in the *Heggie–Hills law*, which states that "hard binaries get harder while soft binaries get softer."

Problem 15.2: Assume a globular cluster consists of stars with mass $m = 0.7 M_\odot$ and has a dispersion velocity of $\langle v \rangle = 20 \,\text{km/s}$. Determine the maximum orbital period of a "hard" binary.

When we include a distribution of masses for the stars in the cluster, the distinction between hard and soft binaries becomes dependent upon the three masses in the encounter. Through numerical experiments a critical velocity is found, such that

$$v_c^2 = \frac{2x(m_1 + m_2 + m_3)}{m_3(m_1 + m_2)}, \qquad (15.13)$$

where m_1 and m_2 are the masses of the binary with an initial relative binding energy x, and m_3 is the mass of the single star. If the velocity of the incoming single star is greater than v_c, then the binary will become softer and may even be disrupted. If the velocity of the incoming single star is less than v_c, then the binary will become harder or may capture the incoming single star. During a capture event, the incoming binary star can be exchanged with one of the original components of the binary system. From simple arguments relating the equipartition of kinetic energy during the encounter, we can see that the star with the lowest mass is most likely to end up with the highest velocity and therefore be ejected from the binary, while the star with the highest mass is most likely to end up with the lowest velocity and therefore be retained in the binary. If the incoming star is the most massive star, then it has a high probability of being exchanged with the lowest mass star in the original binary. These exchange interactions can result in binary systems with components that could not possibly have evolved together in an isolated binary. Two examples of exchange interactions are shown in Fig. 15.1.

Interactions involving two binary systems are even more rare, but occur often enough that they can have an effect on the population of binaries within a cluster. Clearly, the interactions will be much more complicated than three-body interactions, but some qualitative statements can be made about the expected outcomes. For sufficiently distant encounters, each binary will appear to be a single object to the other one. Thus, the tidal dissipation mechanism can be used as an analogy for such an interaction and possible outcomes include the hardening or softening of both binaries, depending on their relative binding energies and relative kinetic energy. If one of the binaries is much harder than the other, then one can

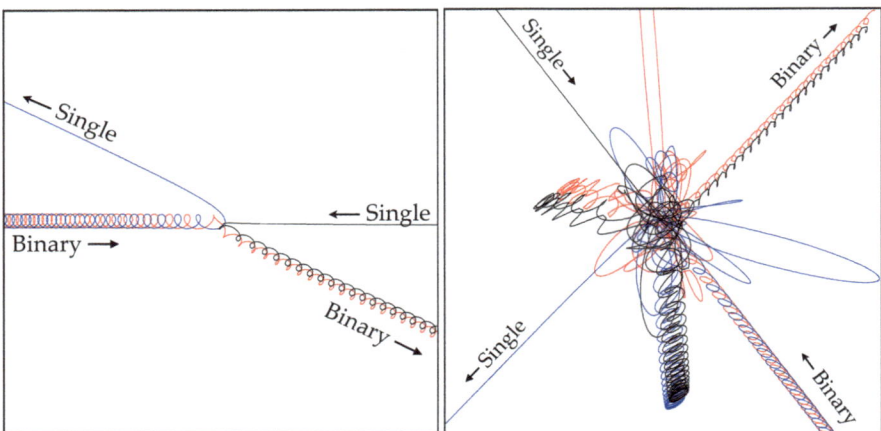

Fig. 15.1 Three-body interactions in which an incoming single star is exchanged with one component of a binary. The incoming single star is shown in *black*, while the components of the initial binary are shown in *blue* and *red*. For the interaction on the *left* is in the plane of the binary and so the interaction is quite simple. For the interaction on the *right*, the incoming single star is coming from an arbitrary direction out of the plane of orbit for the initial binary and the ensuing interaction is very complicated (These simulations were graciously provided by Ladislav Šubr)

use the binary–single interaction as an analogy, treating the hard binary as a single star. This can result in the creation of a hierarchical triple system with a single star in orbit about a tight binary. Particularly close interactions between binaries with comparable properties can even result in the exchange of members so that the binaries leaving the encounter are not the same as those that entered it. Of course, it is also quite probable that one or both of the binaries can be disrupted during the encounter. Numerical simulations have shown that at least one binary is disrupted in ∼88 % of binary–binary encounters.

15.3 N-Body Integration

In order to obtain a quantitative description of few-body encounters, we must numerically integrate and solve the *N*-body problem. In principle, a description of the *N*-body problem is quite simple. Since the timescale for the encounter is extremely short, we ignore any stellar or cluster evolution, and the entire system is described by the gravitational interaction. Thus, the position of each body, $\mathbf{r}_i(t)$, is found from numerically integrating the acceleration:

$$\mathbf{a}_i(t) = -\sum_{j \neq i}^{N} \frac{Gm_j (\mathbf{r}_i - \mathbf{r}_j)}{\left|\mathbf{r}_i - \mathbf{r}_j\right|^3}. \tag{15.14}$$

15.3 N-Body Integration

Since this is a second-order differential equation, the boundary conditions can be the initial positions and velocities ($\mathbf{r}_i(0)$ and $\mathbf{v}_i(0)$). The simplest approach to finding $\mathbf{r}_i(t)$ is to choose a very small time step, Δt, and then compute

$$\mathbf{r}_i(t+\Delta t) = \mathbf{r}_i(t) + \Delta t \mathbf{v}_i(t), \tag{15.15}$$

$$\mathbf{v}_i(t+\Delta t) = \mathbf{v}_i(t) + \Delta t \mathbf{v}_i(t) \tag{15.16}$$

for each star in the system. Although simple in its description, its implementation introduces several complications. The time step must be chosen carefully in order to avoid introducing unacceptable numerical errors. The time step must be small enough that the stars don't move a distance comparable to their separation during a single step. In practice, this is implemented by requiring

$$\Delta t_i = \eta \min\left(\frac{|\mathbf{r}_i - \mathbf{r}_j|}{|\mathbf{v}_i - \mathbf{v}_j|}\right), \tag{15.17}$$

where η is a "fudge factor" such that $\eta \ll 1$. If Δt is too small, an unnecessarily large number of computations are performed and round-off errors begin to grow too large.

Another method for improving the accuracy of the numerical integrations is to introduce a "predictor–corrector" scheme. The expressions in Eqs. (15.15) and (15.16) are simply the leading order terms in a Taylor expansion about t. We can consider extending this expansion to higher order terms, so that

$$\mathbf{r}(t+\Delta t) = \mathbf{r}(t) + \Delta t \mathbf{v}(t) + \frac{1}{2}(\Delta t)^2 \mathbf{a}(t) + \frac{1}{6}(\Delta t)^3 \dot{\mathbf{a}}(t)$$
$$+ \frac{1}{24}(\Delta t)^4 \ddot{\mathbf{a}}(t) + \frac{1}{120}(\Delta t)^5 \dddot{\mathbf{a}}(t), \tag{15.18}$$

$$\mathbf{v}(t+\Delta t) = \mathbf{v}(t) + \Delta t \mathbf{a}(t) + \frac{1}{2}(\Delta t)^2 \dot{\mathbf{a}}(t) + \frac{1}{6}(\Delta t)^3 \ddot{\mathbf{a}}(t)$$
$$+ \frac{1}{24}(\Delta t)^4 \dddot{\mathbf{a}}, \tag{15.19}$$

$$\mathbf{a}(t+\Delta t) = \mathbf{a}(t) + \Delta t \dot{\mathbf{a}}(t) + \frac{1}{2}(\Delta t)^2 \ddot{\mathbf{a}}(t) + \frac{1}{6}(\Delta t)^3 \dddot{\mathbf{a}}(t), \tag{15.20}$$

$$\dot{\mathbf{a}}(t+\Delta t) = \dot{\mathbf{a}}(t) + \Delta t \ddot{\mathbf{a}}(t) + \frac{1}{2}(\Delta t)^2 \dddot{\mathbf{a}}(t), \tag{15.21}$$

where $\ddot{\mathbf{a}}$ and $\dddot{\mathbf{a}}$ are the second and third time derivatives of the acceleration, and we have dropped the subscript i for simplicity. We can compute $\dot{\mathbf{a}}$ directly in terms of the initial conditions $\mathbf{r}$ and $\mathbf{v}$:

$$\dot{\mathbf{a}} = -\sum_{j\neq i}^{N} \frac{Gm_j}{|\mathbf{r}_i - \mathbf{r}_j|^3}\left[(\mathbf{v}_i - \mathbf{v}_j) + 3(\mathbf{r}_i - \mathbf{r}_j)\frac{(\mathbf{r}_i - \mathbf{r}_j)\cdot(\mathbf{v}_i - \mathbf{v}_j)}{|\mathbf{r}_i - \mathbf{r}_j|^2}\right], \tag{15.22}$$

but the higher derivatives of **a** are more problematic. The predictor–corrector scheme works by evaluating Eqs. (15.18) and (15.19) out to the $\dot{a}$ terms, using values calculated directly from the configuration of the system at time t. We then use these values of $\mathbf{r}(t+\Delta t)$ and $\mathbf{v}(t+\Delta t)$ in Eqs. (15.14) and (15.22) to approximate the values of $\mathbf{a}(t+\Delta t)$ and $\dot{\mathbf{a}}(t+\Delta t)$. These predicted values are then used in Eqs. (15.20) and (15.21) to solve for $\ddot{\mathbf{a}}(t)$ and $\dddot{\mathbf{a}}(t)$, which are then inserted into Eqs. (15.18) and (15.19) to correct the values of **r** and **v** out to the $\dddot{\mathbf{a}}$ terms.

> **Problem 15.3:** Solve Eqs. (15.20) and (15.21) to obtain expressions for $(\Delta t)^2 \ddot{\mathbf{a}}(t)$ and $(\Delta t)^3 \dddot{\mathbf{a}}(t)$ in terms of $\mathbf{a}(t)$, $\mathbf{a}(t+\Delta t)$, $\dot{\mathbf{a}}(t)$, and $\dot{\mathbf{a}}(t+\Delta t)$.

15.4 Binary–Cluster Interactions

The interplay between the evolution of the cluster as a whole and the individual dynamical interactions of stars and binaries results in a modification to both the cluster evolution and the binary population. In this section we will look at some of the more prominent features of the combined cluster and binary evolutions. Again, the quantitative description requires numerical solution of the N-body problem, so we will rely on qualitative arguments.

As we have seen in Sect. 14.3, clusters undergo violent relaxation early in their history as the potential changes drastically due to initial mass loss. Following violent relaxation, the equipartition of energy will cause the more massive stars to segregate toward the center of the cluster. Hard binaries will appear to the cluster as single objects with a mass equal to the sum of the components. Therefore, they will also sink toward the center of the cluster. However, since the relaxation time in a globular cluster is on the order of 10^9 year, the more massive stars will have evolved through their life cycle and ended up as either neutron stars or black holes. After a few relaxation times, only stars with masses less than $2 M_\odot$ will still be on the main sequence. Thus, the center of the cluster will be a region with a high density of compact objects and binaries. This will result in an increased number of interactions between binaries and other stars within the center of the cluster. Following the Heggie–Hills law, the soft binaries will become softer or even be disrupted. This process removes energy from the cluster and can accelerate the concentration of stars in the core. On the other hand, the hard binaries will become harder—a process that adds energy to the cluster and delays the core collapse associated with the gravothermal catastrophe. This process of releasing energy from binaries by hardening them is known as *binary burning* and is thought to be responsible for the fact that many globular clusters have not yet undergone core collapse despite being several relaxation times old.

The process of binary burning can be accompanied by exchange interactions. The Galactic globular cluster systems are roughly 10^{10} years old, and so at this age, the average mass of a star in the cluster is $\sim 0.6 M_\odot$. Therefore, typical exchange interactions in the central regions of a cluster will result in the exchange of a compact object such as a black hole, neutron star, or massive white dwarf for one of the low-mass main sequence stars originally in the binary. During the course of the evolution of the cluster, the initial population of binaries will be altered. The overall number of binaries will be reduced as soft binaries are eventually disrupted. As a result, the distribution of orbital periods will be skewed toward shorter periods. This effect will be further enhanced as the hard binaries are further hardened through encounters. Furthermore, the mass distribution of the binary components will be shifted toward higher masses as exchange interactions preferentially remove the low-mass components and replace them with higher mass objects. The resulting population of binary systems consists of a disproportionate number of high-mass compact objects in tight orbits. This is reflected in the proportionately large number of accreting neutron star binaries in the Galactic globular cluster system. It is possible that some of these systems will consist of two objects that evolved completely separate from each other in isolation and were brought together through dynamical interactions.

Problems

15.1. Calculate the average and median values of the eccentricity for a population of binaries with a thermal distribution of eccentricities.

15.2. Assume a globular cluster consists of stars with mass $m = 0.7\,M_\odot$ and has a dispersion velocity of $\langle v \rangle = 20\,\text{km/s}$. Determine the maximum orbital period of a "hard" binary.

15.3. Solve Eqs. (15.20) and (15.21) to obtain expressions for $(\Delta t)^2 \ddot{\mathbf{a}}(t)$ and $(\Delta t)^3 \dddot{\mathbf{a}}(t)$ in terms of $\mathbf{a}(t)$, $\mathbf{a}(t+\Delta t)$, $\dot{\mathbf{a}}(t)$, and $\dot{\mathbf{a}}(t+\Delta t)$.

Appendix A
Useful Constants

A.1 Physical Constants

Table A.1 Physical constants in SI units

Symbol	Constant	Value
c	Speed of light	2.997925×10^8 m/s
e	Elementary charge	1.602191×10^{-19} C
ε_0	Permittivity	8.854×10^{-12} C^{2}s^2/(kg m^3)
μ_0	Permeability	$4\pi \times 10^{-7}$ kg m/C^2
m_H	Atomic mass unit	1.660531×10^{-27} kg
m_e	Electron mass	9.109558×10^{-31} kg
m_p	Proton mass	1.672614×10^{-27} kg
m_n	Neutron mass	1.674920×10^{-27} kg
h	Planck constant	6.626196×10^{-34} J s
$\hbar$	Planck constant	1.054591×10^{-34} J s
$\mathscr{R}$	Gas constant	8.314510×10^3 J/(kg K)
k	Boltzmann constant	1.380622×10^{-23} J/K
σ	Stefan–Boltzmann constant	5.66961×10^{-8} W/(m^2 K^4)
G	Gravitational constant	6.6732×10^{-11} m^3/(kg s^2)

Table A.2 Useful combinations and alternate units

Symbol	Constant	Value
$m_H c^2$	Atomic mass unit	931.50 MeV
$m_e c^2$	Electron rest mass energy	511.00 keV
$m_p c^2$	Proton rest mass energy	938.28 MeV
$m_n c^2$	Neutron rest mass energy	939.57 MeV
h	Planck constant	4.136×10^{-15} eV s
$\hbar$	Planck constant	6.582×10^{-16} eV s
k	Boltzmann constant	8.617×10^{-5} eV/K
hc		1,240 eV nm
$\hbar c$		197.3 eV nm
$e^2/(4\pi\varepsilon_0)$		1.440 eV nm

A.2 Astronomical Constants

Table A.3 Astronomical units

Symbol	Constant	Value
AU	Astronomical unit	$1.4959787066 \times 10^{11}$ m
ly	Light year	$9.460730472 \times 10^{15}$ m
pc	Parsec	2.0624806×10^5 AU
		3.2615638 ly
		3.0856776×10^{16} m
d	Sidereal day	$23^h\ 56^m\ 04.0905309^s$
		$8.61640905309 \times 10^4$ s
d	Solar day	86,400 s
yr	Sidereal year	3.15581498×10^7 s
		365.256308 d
yr	Tropical year	3.155692519×10^7 s
		365.2421897 d
$M_\odot$	Solar mass	1.9891×10^{30} kg
$L_\odot$	Solar luminosity	3.839×10^{26} W
$R_\odot$	Solar radius	6.95508×10^8 m

Appendix B
Atomic Properties of Selected Elements

B.1 Atomic Properties of Selected Elements

Table B.1 Atomic properties of selected elements

Symbol	Element	Mass (m_H)	Ionization energy (eV)	Ground state configuration
^{1}H	Hydrogen	1.007825	13.598	1s
^{2}H		2.014102		
^{3}H∗		3.016049		
^{3}He	Helium	3.016029	24.587	$1s^2$
^{4}He		4.002603		
^{6}He∗		6.018889		
^{6}Li	Lithium	6.015123	5.392	$1s^2 2s$
^{7}Li		7.016004		
^{8}Li∗		8.022487		
^{7}Be∗	Beryllium	7.016930	9.323	$1s^2 2s^2$
^{8}Be∗		8.005305		
^{9}Be		9.012182		
^{10}B	Boron	10.012937	8.298	$1s^2 2s 2p$
^{11}B		11.009305		
^{11}C∗	Carbon	11.011434	11.2603	$1s^2 2s 2p^2$
^{12}C		12.000000		
^{13}C		13.003355		
^{14}C∗		14.003242		
^{12}N∗	Nitrogen	12.018613	14.5341	$1s^2 2s 2p^3$
^{13}N∗		13.005739		
^{14}N		14.003074		
^{15}N		15.000109		

(continued)

Table B.1 (continued)

Symbol	Element	Mass (m_H)	Ionization energy (eV)	Ground state configuration
^{14}O*	Oxygen	14.008596	13.6161	$1s^2 2s^2 p^4$
^{15}O*		15.003066		
^{16}O		15.994915		
^{17}O		16.999132		
^{18}O		17.999161		
^{17}F*	Fluorine	17.002095	17.4228	$1s^2 2s^2 p^5$
^{18}F*		18.000938		
^{18}F		18.998403		
^{20}F*		19.999981		
^{19}Ne*	Neon	19.001880	21.5645	$1s^2 2s^2 p^6$
^{20}Ne		19.992440		
^{21}Ne		20.993847		
^{22}Ne		21.991385		
^{23}Na	Sodium	22.989769	5.1391	[Ne]3s
^{23}Mg*	Magnesium	22.994124	7.6462	[Ne]$3s^2$
^{24}Mg		23.985042		
^{25}Mg		24.985837		
^{26}Mg		25.982593		
^{26}Al*	Aluminum	25.986892	5.9858	[Ne]$3s^2 3p$
^{27}Al		26.981539		
^{28}Si	Silicon	27.976927	8.1517	[Ne]$3s^2 3p^2$
^{29}Si		28.976495		
^{30}Si		29.973770		
^{31}P	Phosphorus	30.973962	10.4867	[Ne]$3s^2 3p^3$
^{32}S	Sulfur	31.972071	10.3600	[Ne]$3s^2 3p^4$
^{33}S		32.971459		
^{34}S		33.967867		
^{35}S*		34.969032		
^{36}S		35.967081		
^{34}Cl*	Chlorine	33.973763	12.9678	[Ne]$3s^2 3p^5$
^{35}Cl		34.968853		
^{36}Cl*		35.968307		
^{37}Cl		36.965903		
^{36}Ar	Argon	35.967545	15.7596	[Ne]$3s^2 3p^6$
^{38}Ar		37.962732		
^{40}Ar		39.962383		
^{41}Ar*		40.964501		
^{41}Ar*		40.964501		
^{38}K*	Potassium	37.969081	4.2407	[Ar]4s
^{39}K		38.963707		
^{40}K*		39.963998		
^{41}K		40.961826		

(continued)

B.1 Atomic Properties of Selected Elements

Table B.1 (continued)

Symbol	Element	Mass (m_H)	Ionization energy (eV)	Ground state configuration
^{40}Ca	Calcium	39.962591	6.1132	[Ar]4s^2
^{41}Ca∗		40.962278		
^{42}Ca		41.958618		
^{43}Ca		42.958767		
^{44}Ca		43.955482		
^{45}Ca∗		44.956187		
^{46}Ca		45.953693		
^{47}Ca∗		46.954546		
^{48}Ca		47.952534		
^{41}Sc∗	Scandium	40.969251	6.5615	[Ar]3d4s^2
^{42}Sc∗		41.965516		
^{45}Sc		44.955912		
^{47}Sc∗		46.952408		
^{48}Sc∗		47.952231		
^{49}Sc∗		48.950024		
^{46}Ti	Titanium	45.952631	6.8281	[Ar]3d^{2}4s^2
^{47}Ti		46.951763		
^{48}Ti		47.947946		
^{49}Ti		48.947870		
^{50}Ti		49.944791		
^{51}V	Vanadium	50.943960	6.7462	[Ar]3d^{3}4s^2
^{50}Cr	Chromium	49.946044	6.7665	[Ar]3d^{5}4s
^{51}Cr∗		50.944767		
^{52}Cr		51.940508		
^{53}Cr		52.940649		
^{54}Cr		53.938880		
^{50}Mn∗	Manganese	49.954238	7.4340	[Ar]3d^{5}4s^2
^{51}Mn∗		50.948211		
^{53}Mn∗		52.941290		
^{55}Mn		54.938045		
^{56}Mn∗		55.938905		
^{57}Mn∗		56.938285		
^{54}Fe	Iron	53.939611	7.9024	[Ar]3d^{6}4s^2
^{55}Fe∗		54.938293		
^{56}Fe		55.934938		
^{57}Fe		56.935394		
^{58}Fe		57.933276		

(continued)

Table B.1 (continued)

Symbol	Element	Mass (m_H)	Ionization energy (eV)	Ground state configuration
^{54}Co*	Cobalt	53.948460	7.8810	[Ar]$3d^7 4s^2$
^{55}Co*		54.941999		
^{57}Co*		56.936291		
^{58}Co*		57.935753		
^{59}Co		58.933915		
^{57}Ni*	Nickel	56.939794	7.6399	[Ar]$3sd^8 4s^2$
^{58}Ni		57.935343		
^{59}Ni*		58.934347		
^{60}Ni		59.930786		
^{61}Ni		60.931056		
^{62}Ni		61.928345		
^{63}Ni*		62.929669		
^{64}Ni		63.927966		

Appendix C
Closest and Brightest Stars

C.1 Closest Stars

Table C.1 Twenty closest stars

Name	Spectral class	Right ascension	Declination	Distance (pc)
V645 centauri	M5.5	$14^h\ 29^m\ 43.0^s$	$-62°\ 40'\ 46''$	1.30
α centauri A	G2	$14^h\ 39^m\ 36.5^s$	$-60°\ 50'\ 02''$	1.34
α centauri B	K1	$14^h\ 39^m\ 35.1^s$	$-60°\ 50'\ 14''$	1.34
Barnard's star	M4	$17^h\ 57^m\ 48.5^s$	$+04°\ 41'\ 36''$	1.83
Wolf 359	M6	$10^h\ 56^m\ 29.2^s$	$+07°\ 00'\ 53''$	2.39
Lalande 21185	M2	$11^h\ 03^m\ 20.2^s$	$+35°\ 58'\ 12''$	2.54
Sirius A	A1	$06^h\ 45^m\ 08.9^s$	$-16°\ 42'\ 58''$	2.63
Sirius B	DA2	$06^h\ 45^m\ 08.9^s$	$-16°\ 42'\ 58''$	2.63
Luyten 726-8A	M5.5	$01^h\ 39^m\ 01.3^s$	$-17°\ 57'\ 01''$	2.68
Luyten 726-8B	M6	$01^h\ 39^m\ 01.3^s$	$-17°\ 57'\ 01''$	2.68
Ross 154	M3.5	$18^h\ 49^m\ 49.4^s$	$-23°\ 50'\ 10''$	2.97
Ross 248	M5.5	$23^h\ 41^m\ 54.7^s$	$+44°\ 10'\ 30''$	3.16
ε eridani	K2	$03^h\ 32^m\ 55.8^s$	$-09°\ 27'\ 30''$	3.23
Lacaille 9352	M1.5	$23^h\ 05^m\ 52.0^s$	$-35°\ 51'\ 11''$	3.29
Ross 128	M4	$11^h\ 47^m\ 44.4^s$	$+00°\ 48'\ 16''$	3.35
EZ aquarii A	M5	$22^h\ 38^m\ 33.4^s$	$-15°\ 18'\ 07''$	3.45
EZ aquarii B	M	$22^h\ 38^m\ 33.4^s$	$-15°\ 18'\ 07''$	3.45
EZ aquarii C	M	$22^h\ 38^m\ 33.4^s$	$-15°\ 18'\ 07''$	3.45
Procyon A	F5	$07^h\ 39^m\ 18.1^s$	$+05°\ 13'\ 30''$	3.50
Procyon B	DA	$07^h\ 39^m\ 18.1^s$	$+05°\ 13'\ 30''$	3.50

C.2 Brightest Stars

Table C.2 Twenty brightest stars

Name	Spectral class	Right ascension	Declination	Apparent magnitude
Sirius A	A1	$06^h\ 45^m\ 08.9^s$	$-16°\ 42'\ 58''$	-1.46
Canopus	F0	$06^h\ 23^m\ 57.1^s$	$-52°\ 41'\ 44''$	-0.72
Arcturus	K1	$14^h\ 15^m\ 39.7^s$	$+19°\ 10'\ 56''$	-0.04
α centauri A	G2	$14^h\ 39^m\ 36.5^s$	$-60°\ 50'\ 02''$	-0.01
Vega	A0	$18^h\ 36^m\ 56.3^s$	$+38°\ 47'\ 01''$	0.03
Rigel	B8	$05^h\ 14^m\ 32.3^s$	$-08°\ 12'\ 06''$	0.12
Procyon A	F5	$07^h\ 39^m\ 18.1^s$	$+05°\ 13'\ 30''$	0.34
Betelgeuse	M2	$05^h\ 55^m\ 10.3^s$	$+07°\ 24'\ 25''$	0.42
α eridani	B3	$01^h\ 37^m\ 42.8^s$	$-57°\ 14'\ 12''$	0.50
β centauri	B1	$14^h\ 03^m\ 49.4^s$	$-60°\ 22'\ 23''$	0.60
Capella A	G8	$05^h\ 16^m\ 41.4^s$	$+45°\ 59'\ 53''$	0.71
Altair	A7	$19^h\ 50^m\ 47.0^s$	$+08°\ 52'\ 06''$	0.77
Aldebaran	K5	$04^h\ 35^m\ 55.2^s$	$+16°\ 33'\ 34''$	0.85
Capella B	G1	$05^h\ 16^m\ 41.4^s$	$+45°\ 59'\ 53''$	0.96
Spica	G1	$13^h\ 25^m\ 11.6^s$	$-11°\ 09'\ 41''$	1.04
Antares	M1	$16^h\ 29^m\ 24.0^s$	$-26°\ 25'\ 55''$	1.04
Pollux	K0	$07^h\ 45^m\ 18.9^s$	$+28°\ 01'\ 34''$	1.15
Fomalhaut	A3	$22^h\ 57^m\ 39.0^s$	$-29°\ 37'\ 20''$	1.16
Deneb	A2	$20^h\ 41^m\ 25.9^s$	$+45°\ 16'\ 49''$	1.25
Mimosa	B0	$12^h\ 47^m\ 43.3^s$	$-59°\ 41'\ 20''$	1.30

Solutions

Problems of Chapter 1

1.1 Use the SIMBAD database at http://simbak.cfa.harvard.edu/simbad/ to determine the position of the star Capella on October 2, 2012. Give the new RA and dec after precessing and then the new values of RA and dec after including proper motion. Note that the proper motions are given as RA=$\mu \sin \phi$ and dec=$\mu \cos \phi$ in this database.

Solution. From SIMBAD we get the following J2000 coordinates: $\alpha = 05^h\ 16^m\ 41.359^s$ and $\delta = +45°\ 59'\ 52.77''$. The proper motion is $\mu_\alpha \cos \delta = 75.25$ mas/year and $\mu_\delta = -426.89$ mas/year. October 2 is the 275th day of the year, so the number of years is

$$N = (2012 - 2000) + 275/365.25 = 12.7529 \text{ year}.$$

The right ascension and declination in degrees are

$$\alpha = 15(5 + 16/60 + 41.359/3600) = 79.1723°,$$

$$\delta = 45 + 59/60 + 52.77/3600 = 45.9980°.$$

The epoch corrections are

$$\Delta \alpha = [3.075 + (1.336) \sin(79.1723) \tan(45.9980)]\, 12.7529 = 56.543\,\text{s},$$

$$\Delta \delta = [20.043 \cos(79.1723)]\, 12.7529 = 48.017''.$$

Therefore, the precessed coordinates are

$$\alpha = 05^h\ 17^m\ 37.902^s,$$

$$\delta = 46°\ 00'\ 40.79''.$$

The proper motions are given in the form $\mu \sin \phi = 0.07525''$/year and $\mu \cos \phi = -0.42689''$/year. Therefore the proper motion changes are

$$\Delta \alpha = \frac{N\mu \sin \phi}{\cos \delta} = \frac{12.7529 \times 0.07525}{\cos 45.9980} = 1.3814'' = 0.0921^{\text{s}},$$

$$\Delta \delta = \frac{N\mu \cos \phi}{=} 12.7529 \times (-0.42689) = -5.444''.$$

Finally, we have the coordinates after accounting for the proper motion:

$$\alpha = 05^{\text{h}} \, 17^{\text{m}} \, 37.994^{\text{s}},$$

$$\delta = 46° \, 00' \, 35.35''.$$

1.2 Given that the luminosity of the sun is $L_\odot = 3.84 \times 10^{26}$ W and the absolute magnitude of the sun is $M = 4.74$, find the apparent magnitude of the sun. The distance to the sun from the earth is $1 \, \text{AU} = 1.496 \times 10^{11}$ m.

Solution.

$$M = m - 5\log_{10}\left(\frac{d}{10\,\text{pc}}\right) \Longrightarrow m = M + 5\log_{10}\left(\frac{d}{10\,\text{pc}}\right).$$

Therefore,

$$m = 4.74 + 5\log_{10}\left(\frac{1.496 \times 10^{11}}{3.086 \times 10^{17}}\right) = -26.83.$$

1.3 The star Sirius is 2.64 pc away from the earth and it has an apparent magnitude of -1.44. What is its luminosity in units of the solar luminosity, $L_\odot$?

Solution.

$$\frac{L}{L_\odot} = \frac{I/4\pi(10\,\text{pc})^2}{I_\odot/4\pi(10\,\text{pc})^2} = 100^{(M-M_\odot)/5}.$$

We find the absolute magnitude of Sirius with

$$M = -1.44 - 5\log_{10} 0.264 = 1.45,$$

so

$$\frac{L}{L_\odot} = 100^{(4.74-1.45)/5} = 20.70.$$

Problems of Chapter 2

2.1 Demonstrate that the orbit lies in a plane by choosing arbitrarily oriented spherical polar coordinates.

Solution. Start with

$$\mathcal{L} = \frac{1}{2}m\left(\dot{r}^2 + r^2\dot{\theta}^2 + r^2\sin^2\theta\,\dot{\phi}^2\right) + \frac{GMm}{r}.$$

Now, the Euler–Lagrange equations read

$$\frac{d}{dt}\frac{\partial \mathcal{L}}{\partial \dot{\phi}} - \frac{\partial \mathcal{L}}{\partial \phi} = \frac{d}{dt}\left(mr^2\sin^2\theta\,\dot{\phi}\right) = 0,$$

$$\frac{d}{dt}\frac{\partial \mathcal{L}}{\partial \dot{\theta}} - \frac{\partial \mathcal{L}}{\partial \theta} = \frac{d}{dt}\left(mr^2\dot{\theta}\right) - mr^2\sin\theta\cos\theta\,\dot{\phi}^2 = 0,$$

$$\frac{d}{dt}\frac{\partial \mathcal{L}}{\partial \dot{r}} - \frac{\partial \mathcal{L}}{\partial r} = \frac{d}{dt}(m\dot{r}) - mr\dot{\theta}^2 - mr\sin^2\theta\,\dot{\phi}^2 + \frac{GMm}{r^2} = 0.$$

When $\theta = \pi/2$, $\sin\theta = 1$ and $\cos\theta = 0$, so these equations read

$$\frac{d}{dt}\left(mr^2\dot{\phi}\right) = 0,$$

$$\frac{d}{dt}\left(mr^2\dot{\theta}\right) = 0,$$

$$\frac{d}{dt}(m\dot{r}) - mr\dot{\theta}^2 - mr\dot{\phi}^2 + \frac{GMm}{r^2} = 0.$$

If, in addition, $\dot{\theta} = 0$, then the θ equation implies that $\dot{\theta}$ is a constant, so $\dot{\theta}$ is always 0 and the equations read

$$\frac{d}{dt}\left(mr^2\dot{\phi}\right) = 0,$$

$$\frac{d}{dt}(0) = 0,$$

$$\frac{d}{dt}(m\dot{r}) - mr\dot{\phi}^2 + \frac{GMm}{r^2} = 0.$$

The middle equation is an identity, and the other two equations are the planar equations.

2.2 Derive Kepler's third law ($GM = a^3\omega^2$) using $J = \mu r^2\dot{\theta}$ and $r = \ell/(1 + e\cos\theta)$.

Solution. First, we notice that the area of an ellipse is

$$A = \pi a b = \pi a^2 \sqrt{1.e^2} = 2 \int_0^\pi \int_0^{r(\theta)} r\,dr\,d\theta = \int_0^\pi r^2\,d\theta.$$

Now,

$$J = \mu r^2 \dot\theta \Longrightarrow \frac{J}{\mu}dt = r^2 d\theta,$$

so

$$\frac{J}{\mu}\int_0^{T/2} dt = \frac{J}{\mu}\frac{T}{2} = \int_0^\pi r^2 d\theta = \pi a^2\sqrt{1-e^2}$$

and

$$\frac{J^2}{\mu^2} = GM\ell = GMa\left(1-e^2\right) = \frac{4\pi^2}{T^2}a^4\left(1-e^2\right) \Longrightarrow GM = \frac{4\pi^2}{T}a^3 = a^3\omega^2.$$

2.3 MT720 is a spectroscopic binary in the Cygnus OB2 association. It is found to have a period of $P = 4.36\,\text{d}$ and an eccentricity of $e = 0.35$. The semi-amplitudes of the radial velocities are $K_1 = 173\,\text{km/s}$ and $K_2 = 242\,\text{km/s}$.

(a) Find $m\sin^3 i$ and $a\sin i$ for each star.
(b) What is the mass ratio of $q = m_2/m_1$?
(c) If $i = 70°$, what are the masses of each star?

Solution. (a) $P = 4.36(86,400) = 3.767 \times 10^5\,\text{s}$, $K_1 = 1.73 \times 10^5\,\text{m/s}$, and $K_2 = 2.42 \times 10^5\,\text{s}$, so

$$m_1 \sin^3 i = \frac{3.767 \times 10^5}{2\pi 6.67 \times 10^{-11}}\left(1 - (0.35)^2\right)^{3/2}\left(4.15 \times 10^5\right)^2 2.42 \times 10^5$$
$$= 3.08 \times 10^{31}\,\text{kg} = 15.4\,\text{M}_\odot,$$

$$m_2 \sin^3 i = \frac{3.767 \times 10^5}{2\pi 6.67 \times 10^{-11}}\left(1 - (0.35)^2\right)^{3/2}\left(4.15 \times 10^5\right)^2 1.73 \times 10^5$$
$$= 2.20 \times 10^{31}\,\text{kg} = 11.0\,\text{M}_\odot,$$

$$a_1 \sin i = \frac{\sqrt{1-(0.35)^2}}{2\pi}\left(1.73 \times 10^5\right)\left(3.767 \times 10^5\right)$$
$$= 9.72 \times 10^9\,\text{m} = 13.96\,\text{R}_\odot,$$

$$a_2 \sin i = \frac{\sqrt{1-(0.35)^2}}{2\pi}\left(2.42 \times 10^5\right)\left(3.767 \times 10^5\right)$$
$$= 1.36 \times 10^{10}\,\text{m} = 19.53\,\text{R}_\odot.$$

(b)
$$q = \frac{m_2}{m_1} = \frac{K_1}{K_2} = \frac{173}{242} = 0.715.$$

(c) If $i = 70°$, then $\sin^3 i = 0.83$ and

$$m_1 = \frac{15.4 \, M_\odot}{0.83} = 18.6 \, M_\odot,$$

$$m_2 = \frac{11.0 \, M_\odot}{0.83} = 13.3 \, M_\odot.$$

Problems of Chapter 3

3.1 GK Vir is an eclipsing spectroscopic binary with an angle of inclination $i = 89.5° \pm 0.6°$, so that it can be considered to be viewed edge on. The orbit is circular with semi-amplitudes of the radial velocities given by $K_1 = 38.6$ km/s and $K_2 = 221.6$ km/s. The orbital period is $P = 0.344$ d. The time required for the light curve to drop to its lowest value is $t_b - t_a = 89.6$ s, while the time required for the light curve to begin rising again is $t_c - t_a = 817$ s. Use this information to find:

(a) The radii of both stars
(b) The masses of both stars

Solution. (a) If the orbit is circular, then $K_1 = v_1$ and $K_2 = v_2$. Therefore the radii are found from

$$r_1 = \frac{K_1 + K_2}{2}(t_b - t_a) = \frac{2.602 \times 10^5 \, \text{m/s}}{2}(89.6 \, \text{s}) = 1.166 \times 10^7 \, \text{m} = 0.017 \, R_\odot,$$

$$r_2 = \frac{K_1 + K_2}{2}(t_c - t_a) = \frac{2.602 \times 10^5 \, \text{m/s}}{2}(817 \, \text{s}) = 1.063 \times 10^8 \, \text{m} = 0.153 \, R_\odot.$$

(b) Use the spectroscopic binary formula to find the masses:

$$m_1 = \frac{P}{2\pi G}(K_1 + K_2)^2 K_2 = \frac{(0.344)(86,400)}{2\pi(6.67 \times 10^{-11})}\left(2.602 \times 10^5\right)^2 \left(2.216 \times 10^5\right)$$

$$= 1.064 \times 10^{30} \, \text{kg} = 0.532 \, M_\odot,$$

$$m_2 = \frac{P}{2\pi G}(K_1 + K_2)^2 K_1 = \frac{(0.344)(86,400)}{2\pi(6.67 \times 10^{-11})}\left(2.602 \times 10^5\right)^2 \left(3.86 \times 10^4\right)$$

$$= 1.853 \times 10^{29} \, \text{kg} = 0.093 \, M_\odot.$$

3.2 For a gas of neutral hydrogen atoms, at what temperature is the number of atoms in the first excited state only 1% of the number of atoms in the ground state? At what temperature is the number of atoms in the first excited state 10% of the number of atoms in the ground state?

Solution.

$$\frac{N_2}{N_1} = r = \frac{8}{2}e^{-3e_0/4kT} \implies T = \frac{3E_0}{4k\ln(4/r)}.$$

Use $E_0 = 13.6\,\text{eV}$ and $k = 8.617 \times 10^{-5}\,\text{eV/K}$, so

$$r = 0.01 \implies T = 19{,}756\,\text{K}$$

and

$$r = 0.1 \implies T = 32{,}088\,\text{K}.$$

3.3 A typical atmosphere found on a white dwarf of spectral type DB is pure helium. The ionization energies of neutral helium and singly ionized helium are $\xi_I = 24.6\,\text{eV}$ and $\xi_{II} = 54.4\,\text{eV}$, respectively. The partition functions are $Z_I = 1$, $Z_{II} = 2$, and $Z_{III} = 1$. Use $P_e = 20\,\text{N/m}^2$ for the electron pressure.

(a) Use the Saha equation to find N_{II}/N_I and N_{III}/N_{II} for temperatures of 5,000 K, 15,000 K, and 25,000 K.
(b) Show that $N_{II}/N_{\text{total}} = N_{II}/(N_I + N_{II} + N_{III})$ can be expressed in terms of the ratios N_{II}/N_I and N_{III}/N_{II}.
(c) Plot N_{II}/N_{total} for temperatures between 5,000 K and 25,000 K. What is the temperature for which $N_{II}/N_{\text{total}} = 0.5$?

Solution. (a) From the Saha equation, we have

$$\frac{N_{II}}{N_I} = \left(\frac{2k}{P_e}\right)\left(\frac{2\pi m_e c^2 k}{(hc)^2}\right)^{3/2} \frac{Z_{II}}{Z_I} T^{5/2} e^{-\xi_I/kT}.$$

Some useful constants to know are $m_e c^2 = 511\,\text{keV}$, $hc = 1{,}240\,\text{eV\,nm}$, $k = 8.617 \times 10^{-5}\,\text{eV/K}$, and $1\,\text{Pa} = 6.242 \times 10^{18}\,\text{eV/m}^3$. Therefore,

$$\frac{N_{II}}{N_I} = \frac{2(8.617 \times 10^{-5}\,\text{eV/K})}{20(6.242 \times 10^{18}\,\text{eV/m}^3)}\left(\frac{2\pi(5.11 \times 10^5\,\text{eV})(8.617 \times 10^{-5}\,\text{eV/K})}{(1.240 \times 10^{-6}\,\text{eV\,m})^2}\right)^{3/2}$$

$$\times \frac{2}{1} T^{5/2} e^{(24.6\,\text{eV})/(T 8.617 \times 10^{-5}\,\text{eV/K})}$$

$$= 6.66 \times 10^{-3} \left(\frac{T}{\text{K}}\right)^{5/2} e^{-(2.85 \times 10^5\,\text{K})/T}.$$

So

$$T = 5{,}000\,\text{K} \implies 1.88 \times 10^{-18},$$
$$T = 15{,}000\,\text{K} \implies 9.96 \times 10^{-1},$$
$$T = 25{,}000\,\text{K} \implies 7.23 \times 10^3.$$

Solutions

For N_{III}/N_{II} the only thing that changes is $\xi = 54.4\,\text{eV}$ and $Z_{III} = 1$, so

$$\frac{N_{III}}{N_{II}} = 1.67 \times 10^6 \left(\frac{T}{K}\right)^{5/2} e^{-(6.31 \times 10^5\,\text{K})/T}.$$

So

$$T = 5{,}000\,\text{K} \Longrightarrow 4.31 \times 10^{-49},$$
$$T = 15{,}000\,\text{K} \Longrightarrow 2.42 \times 10^{-11},$$
$$T = 25{,}000\,\text{K} \Longrightarrow 1.78 \times 10^{-3}.$$

(b)
$$\frac{N_{II}}{N_{\text{total}}} = \frac{N_{II}}{N_I + N_{II} + N_{III}} = \frac{1}{\frac{N_I}{N_{II}} + 1 + \frac{N_{III}}{N_{II}}}.$$

(c) From the plot, we see that $15{,}000 < T < 16275$.

Problems of Chapter 4

4.1 Using the density distribution

$$\frac{M}{4R^2} \frac{\sin(\pi r/R)}{r},$$

compute $m(r)$.

Solution.

$$m(r) = 4\pi \int_0^r \rho(r) r^2 dr = \frac{\pi M}{R^2} \int_0^r r \sin(\pi r/R) dr$$
$$= \frac{M}{\pi} \left[\sin(\pi r/R) - \frac{\pi r}{R} \cos(\pi r/R)\right].$$

4.2 Using the density from Problem 4.1, compute Ω and show that $\alpha = 0.75$.

Solution.

$$\Omega = -\int_0^M \frac{Gm\,dm}{r} = -\frac{GM^2}{R^2} \int_0^R \left[\sin(\pi r/R) - \frac{\pi r}{R}\cos(\pi r/R)\right] \sin(\pi r/R) dr.$$

Make the substitution $x = \pi r/R$, so that

$$\Omega = -\frac{GM^2}{\pi R}\left[\int_0^\pi \sin^2 x\,dx - \int_0^\pi x\cos x\sin x\,dx\right]$$

$$= -\frac{GM^2}{\pi R}\left[\frac{1}{2}(x - \sin x\cos x) - \frac{1}{8}(\sin 2x - 2x\cos 2x)\right]_0^\pi$$

$$= -\frac{3GM^2}{4R}.$$

So $\alpha = 0.75$.

4.3 Assuming that a star of mass M is devoid of nuclear energy sources, determine its radius as a function of time if it maintains a constant luminosity L. Assume that the star is in hydrostatic equilibrium.

Solution. Assume that the star is in hydrostatic equilibrium and that the virial theorem holds. Therefore,

$$L = \dot{E} = \dot{K} + \dot{\Omega} + \dot{U} = 0 + \dot{\Omega} - \frac{1}{2}\dot{\Omega} = -\frac{d}{dt}\left(\frac{1}{2}\frac{\alpha GM^2}{R}\right) = -\frac{\alpha GM^2}{2R^2}\frac{dR}{dt}.$$

Consequently,

$$\frac{dR}{dt} = -\frac{2LR^2}{\alpha GM^2}.$$

Note that we can find the time dependence of R by integrating

$$\int_{R_0}^R \frac{dR}{R^2} = -\int_0^t \frac{2L}{\alpha GM^2}\,dt.$$

So

$$\frac{-1}{R} + \frac{1}{R_0} = -\frac{2Lt}{\alpha GM^2} \Longrightarrow R = \frac{\alpha R_0 GM^2}{\alpha GM^2 + 2LR_0 t}.$$

Problems of Chapter 5

5.1 Evaluate the pressure integral using the Maxwell–Boltzmann distribution and show that you get the ideal gas law.

Solution. First, we note that the Maxwell–Boltzmann distribution has the following property:

$$n = \int_0^\infty n(p)\,dp = \frac{4\pi n}{(2\pi mkT)^{3/2}}\int_0^\infty p^2 e^{-p^2/2mkT}\,dp.$$

Now, using $v = p/m$, the pressure integral gives

Solutions

$$P = \frac{1}{3}\int_0^\infty \frac{4\pi n p^4}{m(2\pi mkT)^{3/2}} e^{-p^2/2mkT}\,dp.$$

Use the fact that

$$pe^{-p^2/2mkT} = -mkT \frac{d}{dp} e^{-p^2/2mkT},$$

and integrate by parts to get

$$P = \frac{1}{3}\frac{4\pi n}{m(2\pi mkT)^{3/2}} mkT \int_0^\infty 3p^2 e^{-p^2/2mkT}\,dp$$

or

$$P = kT \frac{4\pi n}{(2\pi mkT)^{3/2}} \int_0^\infty p^2 e^{-p^2/2mkT}\,dp = nkT.$$

5.2 The relativistic form of the kinetic energy is $K = (\gamma - 1)mc^2$. Determine the density for which $\gamma \sim 1.1$. (i.e., $p^2/2m \sim 0.1\,mc^2$).

Solution.

$$\bar{E} = \frac{3}{5}E_F = \frac{3}{5}\frac{\hbar^2}{2m_e}\left[3\pi^2\left(\frac{\mathscr{Z}}{\mathscr{A}}\right)\frac{\rho}{m_H}\right]^{2/3} = 0.1 m_e c^2.$$

Some algebra gives

$$\rho = \left(\frac{\mathscr{A}}{\mathscr{Z}}\right)\frac{m_H}{\pi^2}\frac{1}{3^{5/2}}\left[\frac{m_e c}{\hbar}\right]^3 = \frac{16\pi}{3^{5/2}} m_H \left[\frac{m_e c}{h}\right]^3.$$

Taking $\mathscr{Z}/\mathscr{A} = 1/2$ using 1.661×10^{-27} kg for m_H and 2.426×10^{-12} m for the Compton wavelength $(h/m_e c)$, we get

$$\rho = 3.7 \times 10^8\,\text{kg/m}^3.$$

5.3 For highly relativistic electrons, $E \simeq pc$ where $p = \sqrt{p_x^2 + p_y^2 + p_z^2}$. Following Eq. (5.28), we have $E = \hbar c \frac{\pi}{L}\sqrt{N_x^2 + N_y^2 + N_z^2} = \hbar c \pi N/L$. Derive Eq. (5.45) as follows:

(a) The number of states is related to N by $N_s = \left(\frac{\pi}{3}N^3\right)$. Use this to write E as a function of N_s and show that

$$g(E) = \frac{dN_s}{dE} = \frac{1}{\pi^2}\left(\frac{L}{\hbar c}\right)^3 E^2.$$

Solution.

$$E = \left(\frac{\hbar c \pi}{L}\right)\left(\frac{3N_e}{\pi}\right)^{1/3} \implies N_e = E^3 \left(\frac{L}{\hbar c \pi}\right)^3 \left(\frac{\pi}{3}\right),$$

so

$$g(E) = \frac{dN_e}{dE} = 3E^2 \left(\frac{L}{\hbar c \pi}\right)^3 \left(\frac{\pi}{3}\right) = \frac{1}{\pi^2}\left(\frac{L}{\hbar c}\right)^3 E^2.$$

(b) Define $E_F = E(N_e)$ where N_e is the total number of free electrons in the star and show that

$$\bar{E} = \frac{1}{N_e} \int_0^{E_F} g(E) E \, dE = \frac{3}{4} E_F.$$

Solution.

$$\bar{E} = \frac{1}{N_e} \int_0^{E_F} \frac{1}{\pi^2}\left(\frac{L}{\hbar c}\right)^3 E^3 dE = \frac{1}{N_e} \frac{1}{\pi^2} \left(\frac{L}{\hbar c}\right)^3 \frac{1}{4} E_F^4.$$

From part (a), we have

$$N_e = E_F^3 \left(\frac{L}{\hbar c \pi}\right)^3 \left(\frac{\pi}{3}\right),$$

so

$$\bar{E} = \left(\frac{\hbar c \pi}{L}\right)^3 \left(\frac{3}{\pi}\right) \frac{1}{\pi^2} \left(\frac{L}{\hbar c}\right)^3 \frac{1}{4} E_F = \frac{3}{4} E_F.$$

(c) Noting that $L = V^{1/3}$, show that

$$P = -\frac{\partial}{\partial V}(N_e \bar{E}) = \frac{1}{4} \hbar c \left(3\pi^2\right)^{1/3} n_e^{4/3},$$

where $n_e = N_e/V$.

Solution.

$$P = -\frac{\partial}{\partial V}\left(N_e \frac{3}{4} \left(\frac{\hbar c \pi}{L}\right) \left(\frac{3N_e}{\pi}\right)^{1/3}\right) = -\frac{3}{4} \hbar c \left(3\pi^2\right)^{1/3} N_e^{4/3} \frac{\partial}{\partial V} V^{-1/3}$$

$$= \frac{1}{4} \hbar c \left(3\pi^2\right)^{1/3} N_e^{4/3} V^{-4/3} = \frac{1}{4} \hbar c \left(3\pi^2\right)^{1/3} n_e^{4/3}.$$

5.4 Do this integral (Eq. (265)) for a classical degenerate gas to obtain Eq. (268).

Solution. Using $N = 2Lp/h$, the number of momentum states per volume is given by

$$n(p) d(p) = \frac{2}{8V} 4\pi N^2 dN = \frac{8\pi}{h^3} p^2 dp,$$

so

$$u = \frac{1}{\rho} \int_0^{p_F} n(p) \varepsilon(p) dp = \frac{1}{\rho} \int_0^{p_F} \frac{8\pi}{h^3} p^2 \frac{p^2}{2m} dp = \frac{4\pi}{mh^3 \rho} \frac{1}{5} p_F^5.$$

Now, we need to rewrite p_F in terms of P, so

$$p_F = \sqrt{2mE} = \sqrt{2m}\frac{\hbar}{\sqrt{2m}}\left[(3\pi^2)\left(\frac{\mathscr{Z}}{\mathscr{A}}\right)\frac{\rho}{m_H}\right]^{1/3}$$

and

$$P = \frac{1}{5}\frac{\hbar^2(3\pi^2)^{2/3}}{m}\left[\frac{\mathscr{Z}}{\mathscr{A}}\frac{\rho}{m_H}\right] \Longrightarrow \left[\frac{\mathscr{Z}}{\mathscr{A}}\frac{\rho}{m}\right]^{1/3} = \left[\frac{5Pm}{\hbar^2(3\pi^2)^{2/3}}\right]^{1/5},$$

so

$$p_F = \hbar^{3/5}(3\pi^2)^{1/5}(5Pm)^{1/5}$$

and

$$u = \frac{4\pi}{mh^3\rho}\frac{1}{5}\hbar^3(3\pi^2)5Pm = \frac{3}{2}\frac{P}{\rho}.$$

Problems of Chapter 6

6.1 Show that you can recover the expression for P_{rad} in Eq. (5.27) if you use B_λ, the blackbody expression for the specific intensity.

Solution. The integral is

$$P_{\text{rad}} = \frac{1}{c}\int_0^\infty d\lambda \int_0^{2\pi} d\phi \int_0^\pi d\theta I_\lambda \cos^2\theta \sin\theta$$

and the specific intensity for a blackbody is

$$B_\lambda = \frac{2hc^2}{\lambda^5(e^{hc/\lambda kT}-1)}.$$

Since B_λ is independent of θ, we can quickly do the angle integrals, so

$$P_{\text{rad}} = \frac{8\pi hc}{3}\int_0^\infty \frac{d\lambda}{\lambda^5(e^{hc/\lambda kT}-1)}.$$

Now, make the variable substitution:

$$y = \frac{hc}{kT}\frac{1}{\lambda},$$

so
$$P_{\text{rad}} = \frac{8\pi k^4 T^4}{3h^3 c^3} \int_0^\infty \frac{y^3 dy}{(e^y - 1)}.$$

Looking at the integral, we note that it can be rewritten:
$$\int_0^\infty \frac{y^3 dy}{(e^y - 1)} = \int_0^\infty y^3 e^{-y} \left(\frac{1}{1 - e^{-y}}\right) dy,$$

and the term in brackets is simply the geometric series:
$$\frac{1}{1 - e^{-y}} = \sum_{n=0}^\infty e^{-ny},$$

so
$$\int_0^\infty \frac{y^3 dy}{(e^y - 1)} = \sum_{n=1}^\infty \int_0^\infty y^3 e^{-ny} dy.$$

Make another variable substitution $z = ny$, and the integral becomes
$$\sum_{n=1}^\infty n^{-4} \int_0^\infty z^3 e^{-z} dz = \sum_{n=1}^\infty n^{-4} \Gamma(4) = \zeta(4)\Gamma(4).$$

Finally, $\Gamma(4) = 3!$ and $\zeta(4) = \pi^4/90$, so
$$\int_0^\infty \frac{y^3 dy}{(e^y - 1)} = \frac{\pi^4}{15}$$

and
$$P_{\text{rad}} = \left(\frac{8\pi^5 k^4}{15 h^3 c^3}\right) \frac{1}{3} T^4.$$

6.2 The relativistic momentum of a particle is $\gamma m v$ and the relativistic energy is $\gamma m c^2$, where $\gamma = (1 - v^2/c^2)^{-1/2}$. Use the relativistic conservation of momentum and energy to show that a free particle cannot absorb a photon.

Solution. Start with the particle at rest and an incoming photon and assume that both energy and momentum are conserved:
$$\frac{hf}{c} = \gamma m v,$$
$$hf + mc^2 = \gamma m c^2.$$

If momentum is conserved, then $hf = \gamma m v c$. Substitute this into the energy equation to find
$$mc^2 + \gamma m v c = \gamma m c^2.$$

Dividing out the m and solving for the final velocity of the particle, we find that $v = c$. Since no particle can travel at the speed of light, then the assumption that energy and momentum are both conserved is false. Therefore a free particle cannot absorb a photon.

6.3 Using Eq. (6.18) and setting $\tau_r = 0$, the specific intensity at the surface is given by

$$I_\lambda = -\frac{1}{\cos\theta} \int_0^\infty S_\lambda(t,\theta) e^{-t/\cos\theta} dt.$$

Show that if $S_\lambda(\tau_r) = a_\lambda + b_\lambda \tau$, then $I_\lambda = a_\lambda + b_\lambda \cos\theta$.

Solution.

$$I_\lambda = -\frac{1}{\cos\theta} \int_0^\infty (a_\lambda + b_\lambda t) e^{-t/\cos\theta} dt$$

$$= -\frac{a_\lambda}{\cos\theta} \int_0^\infty e^{-t/\cos\theta} dt - \frac{b_\lambda}{\cos\theta} \int_0^\infty t e^{-t/\cos\theta} dt$$

$$= a_\lambda e^{-t/\cos\theta} \Big|_0^\infty + b_\lambda e^{-t/\cos\theta} (t + \cos\theta) \Big|_0^\infty$$

$$= a_\lambda + b_\lambda \cos\theta.$$

Problems of Chapter 7

7.1 The source of the luminosity of the sun is the fusion of four ^{1}H nuclei to form one ^{4}He nucleus. The energy from each reaction comes from the conversion of mass to energy via $E = mc^2$. Using the luminosity of the sun, calculate the mass loss rate (in kg/s) due to nuclear fusion.

Solution.

$$L = \frac{dE}{dt} = \frac{dm}{dt}c^2 \Longrightarrow \frac{dm}{dt} = \frac{L}{c^2} = \frac{3.84 \times 10^{26}}{9 \times 10^{16}} = 4.27 \times 10^9 \,\text{kg/s}.$$

7.2 In the p-p chain, the net energy released for each ^{4}He created is 26 MeV. Assuming that the entire luminosity of the sun is generated by the p-p chain, calculate the mass of ^{1}H that is converted to ^{4}He (in kg/s).

Solution. The number of reactions per second is $N = L/\Delta E$ where $\Delta E = 26\,\text{MeV}$ and the mass of hydrogen consumed per second is $dm_H/dt = 4Nm_H$. Therefore,

$$\frac{dm_H}{dt} = 4\frac{L}{\Delta E}m_H = 4\frac{Lm_Hc^2}{\Delta E c^2} = 4\frac{L}{c^2}\frac{m_Hc^2}{\Delta E}$$

$$= 4\left(4.27 \times 10^9 \,\text{kg/s}\right)\frac{938.27\,\text{MeV}}{26\,\text{MeV}} = 6.15 \times 10^{11}\,\text{kg/s}.$$

7.3 In a fully convective core, the material within the core is continually mixed so that the mass fractions are uniform throughout the core. A star has a fully convective core that is pure ^{1}H and has a mass of $0.1\,M_\odot$.

(a) Assume that all of the hydrogen is burned to produce a core of pure ^{4}He. What is the mass of the resulting helium core? What is the total amount of energy released during the hydrogen-burning phase?

Solution. The number of ^{4}He atoms in the helium core is four times the number of ^{1}H atoms in the original core. Taking the atomic mass of ^{1}H to be $1.007825\,m_H$ and the atomic mass of ^{4}He to be $4.0026032\,m_H$, the mass of the helium core is

$$M_{He} = M_H \frac{4.0026032}{4 \times 1.007825} = 0.09929\,M_\odot.$$

The total amount of energy released is $(M_H - M_{He})c^2 = 1.281 \times 10^{44}\,\text{J}$.

(b) Assume that all of the helium in the helium core is subsequently burned to produce a core of pure ^{12}C. What is the mass of the resulting carbon core? What is the total amount of energy released during the helium-burning phase?

Solution. Using the same procedure as part (a) but taking the atomic mass of ^{12}C $= 12\,m_H$, the mass of the carbon core is

$$M_C = M_{He} \frac{12}{3 \times 4.0026032} = 0.09922\,M_\odot.$$

The total amount of energy released is $(M_{He} - M_C)c^2 = 1.162 \times 10^{43}\,\text{J}$.

Problems of Chapter 8

8.1 Solve the Lane–Emden equation for $n = 1$. Calculate the mass of the star in terms of the central density ρ_c.

Solution. For $n = 1$, the Lane–Emden equation becomes

$$\xi^2 \Theta'' + 2\xi \Theta' + \xi^2 \Theta = 0,$$

which is the differential equation for a spherical Bessel function of order 0. Therefore, the general solution is

$$\Theta = A j_0(\xi) + B y_0(\xi) = A \frac{\sin \xi}{\xi} - B \frac{\cos \xi}{\xi}.$$

Imposing boundary conditions $\Theta(0) = 1$ and $\Theta'(0) = 0$ gives

$$\Theta(\xi) = \frac{\sin \xi}{\xi}.$$

The first zero is found at $\xi_1 = \pi$, and so the mass of the star is

$$M = -4\pi \alpha^3 \rho_c \xi_1^2 \left(\frac{\cos \xi_1}{\xi_1} - \frac{\sin \xi_1}{\xi_1^2} \right) = 4\pi^2 \alpha^3 \rho_c.$$

8.2 The solar value of the mean molecular weight is 0.61. Use this to determine β for the Eddington Standard Model of the sun.

Solution. The solar value for the mass is $M = M_\odot$, so we numerically solve the quartic equation: $(0.00415)\beta^4 + \beta - 1 = 0$. This gives $\beta \simeq 0.996$. This means that the pressure in the sun is almost entirely due to gas pressure.

8.3 Using the value of β found in Problem 8.2 and the solutions for the $n = 3$ polytrope, determine the central density and central temperature of the sun in the Eddington Standard Model.

Solution. Using $M = M_\odot = 2 \times 10^{30}$ kg and $R = R_\odot = 6.96 \times 10^8$ m, the average density is

$$\bar{\rho} = \frac{3M}{4\pi R^3} = 1.416 \times 10^3 \, \text{kg/m}^3.$$

From Table 8.1, we have $D_3 = 54.21$, and so

$$\rho_c = D_3 \bar{\rho} = 7.677 \times 10^4 \, \text{kg/m}^3.$$

The temperature is found from

$$T_c = \left[\frac{3\mathscr{R}(1-\beta)}{a\mu\beta} \right]^{1/3} \rho_c^{1/3}.$$

Using $\mathscr{R} = 8{,}314.51$ J/kgK, $a = 7.5646 \times 10^{-16}$ J/m³K⁴, $\mu = 0.61$, and $\beta = 0.996$, this gives

$$T_c = 2.55 \times 10^7 \, \text{K}.$$

8.4 Numerically solve the $n = 3$ polytrope equation using solar values to determine the radius at which the temperature is high enough for the p-p chain to be operating ($T_6 = 15$).

Solution. We want to find the value of ξ for which $T = 1.5 \times 10^7$ K. Note that

$$T = \left[\frac{3\mathscr{R}(1-\beta)}{a\mu\beta} \right]^{1/3} \rho^{1/3} = \left[\frac{3\mathscr{R}(1-\beta)}{a\mu\beta} \right]^{1/3} \left(\rho_c \Theta^3(\xi) \right)^{1/3} = T_c \Theta(\xi).$$

Therefore, we numerically solve the Lane–Emden equation and determine the value of ξ for which $\theta = T/T_c = 15/25 = 0.6$. This is

$$\xi = 1.935.$$

Now we want to convert this into solar radii. Noting that $R_\odot = \alpha R_3$, we find that

$$R = \frac{R_\odot}{R_3}\xi = \frac{1.935}{6.90}R_\odot = 0.28R_\odot = 1.95 \times 10^8 \,\text{m}.$$

8.5 Using $\nu = 3.5$, determine the upper bound on the mass of a main sequence star.

Solution. Use solar units for the mass-luminosity relation, then the proportionality constant is 1, and the maximum mass occurs when the luminosity reaches the Eddington luminosity. For massive stars, the dominant source of opacity is electron scattering; therefore,

$$L_{\text{Edd}} = M_{\text{max}}^\nu = 3.2 \times 10^4 M_{\text{max}} \implies M_{\text{max}} = \left(3.2 \times 10^4\right)^{1/(\nu-1)} = 63.4\,M_\odot.$$

Problems of Chapter 9

9.1 A shell of hydrogen is burning around a helium core of radius r. If the gas is described by an ideal gas, what is the minimum outer radius of this shell if the hydrogen burning is stable? What is the minimum outer radius if the gas is a non-relativistic degenerate gas?

Solution. For an ideal gas, $a = 1$, and so the minimum thickness of the shell is given by $4\ell = r$. Therefore, the outer radius of the shell is $5r/4$. If the gas is a nonrelativistic degenerate gas, then $a = 5/3$, and then $4\ell = 5r/3$ and so $\ell = 5r/12$ and the outer radius is at $17r/12$.

9.2 For an ideal gas, what is the stable oscillation frequency for $m = 1\,M_\odot$ and $r_0 = 1\,R_\odot$?

Solution. For an ideal gas, $\gamma_a = 5/3$. Therefore,

$$\omega = \sqrt{Gm/r_0^3} = \sqrt{\frac{(6.67 \times 10^{-11})(2 \times 10^{30})}{(6.96 \times 10^8)^3}} = 6.29 \times 10{-4}\,\text{Hz}.$$

This corresponds to an oscillation frequency of 10^{-4} Hz and a period of oscillation of about 10,000 s, or about 2.8 h.

9.3 Starting with Eq. (9.44), show that the condition for convection can also be written
$$\frac{d\ln P}{d\ln T} < \frac{\gamma_a}{\gamma_a - 1}.$$

Solution. The condition for convection is
$$\frac{dT}{dr} > \frac{T}{P}\left(1 - \frac{1}{\gamma_a}\right)\frac{dP}{dr}.$$

Dividing by T and rearranging slightly gives
$$\frac{1}{T}\frac{dT}{dr} > \frac{1}{P}\frac{dP}{dr}\left(\frac{\gamma_a - 1}{\gamma_a}\right).$$

Noting that
$$\frac{1}{T}\frac{dT}{dr} = \frac{d\ln T}{dr},$$
we have
$$\frac{d\ln T}{dr}\frac{dr}{d\ln P} > \frac{\gamma_a - 1}{\gamma_a}$$
or
$$\frac{d\ln P}{d\ln T} < \frac{\gamma_a}{\gamma_a - 1}.$$

Problems of Chapter 10

10.1 The Jeans length (R_J) is defined to be the minimum radius necessary to collapse a cloud of density ρ_0.

(a) Use the expression of the Jeans mass to obtain one for the Jeans length.

Solution. The Jeans length is related to the Jeans mass by
$$\rho_0 = \frac{3M_J}{4\pi R_J^3} \implies R_J = \left(\frac{3M_J}{4\pi\rho_0}\right)^{1/2},$$
so
$$R_J = \left(\frac{3}{4\pi\rho_0}\left[\frac{5\mathscr{R}T}{\mu G}\right]^{3/2}\left[\frac{3}{4\pi\rho_0}\right]^{1/2}\right)^{1/3} = \sqrt{\frac{15\mathscr{R}T}{4\pi\mu G\rho_0}}.$$

(b) For a typical diffuse hydrogen cloud, $T = 50\,\text{K}$, and $n = 5 \times 10^8\,\text{m}^{-3}$. If we assume that the cloud is entirely composed of H I, $\rho_0 = m_H n = 8.4 \times 10^{-19}\,\text{kg/m}^3$. Taking $\mu = 1$, determine R_J for this cloud.

Solution. Using the following constants:

$$\mathscr{R} = 8.314 \times 10^3 \, \text{J}\,\text{kg}^{-1}\,\text{K}^{-1},$$
$$G = 6.67 \times 10^{-11} \, \text{m}^3\,\text{kg}^{-1}\,\text{s}^{-2},$$

and so

$$R_J = \sqrt{\frac{15\,(8.314 \times 10^3)\,50}{4\pi\,(6.67 \times 10^{-11})\,(8.4 \times 10^{-19})}}$$
$$= 9.41 \times 10^{16}\,\text{m} = 1.37 \times 10^7\,R_\odot.$$

10.2 Using the opacity for dust and taking the initial density to be $\rho \simeq 10^{-10}\,\text{kg/m}^3$, determine the radius of the star when the optical depth becomes $2/3$ at the center. Assume constant density.

Solution. Assuming constant density, then

$$R = \frac{\tau}{\kappa \rho} = \frac{2}{3 \times 10^{-3} \times 10^{-10}} = 6.7 \times 10^{12}\,\text{m} = 9.6 \times 10^3\,R_\odot.$$

10.3 The typical metal composition of a star with solar metallicity is described by $X = 0.68$, $Y = 0.3$, and $Z = 0.02$. Using these values, determine the radius and central temperature of the star after ionization of hydrogen and helium.

Solution. Using the mass fractions, the energy needed to ionize the hydrogen and helium is

$$E = \frac{M}{4m_H}\left[0.68\,(4\,(13.6\,\text{eV})+2\,(4.48\,\text{e}))+3\,(78.98\,\text{eV})\right] = \frac{M}{4m_H}66.78\,\text{eV}.$$

Converting to Joules gives

$$E = (3.2 \times 10^{39}\,\text{J})\frac{M}{M_\odot}.$$

Equating this energy to the energy released from collapse gives

$$\frac{R}{R_\odot} = \frac{GM^2}{2ER_\odot} = 60\frac{M}{M_\odot}.$$

Thus, the addition of metallicity does not change the results.

10.4 Assume that the opacity is dominated by the H^- ion. Show that

$$R \propto \left(\frac{M}{M_\odot}\right)^{13/17} R_\odot,$$

when a star enters the Henyey track. If a 1 M$_\odot$ star has a radius of 5 R$_\odot$ when it enters the Henyey track, what is the radius of a 2 M$_\odot$ star at this point in its evolution?

Solution. When the star enters the Henyey track, $M^{s-n+3} \propto R^{s-3n+2}$. For H$^-$ opacity, $s = -9$ and $n = 1/2$, so $M^{13} \propto R^{17}$. Therefore,

$$R \propto \left(\frac{M}{M_\odot}\right)^{13/17} R_\odot.$$

If a 1 M$_\odot$ star has a radius of 5 R$_\odot$, then the proportionality constant is 5, and so a 2 M$_\odot$ star has a radius of

$$R = 5^{13/17} R_\odot = 8.5 R_\odot.$$

Problems of Chapter 11

11.1 Assume that the core contraction associated with the increase in μ is homologous, so that throughout the core, $r \to r + \delta r$. For an energy rate given by $q = q_0 \rho T^\beta$, show that the fractional change in energy rate is given by

$$\frac{\delta q}{q} = -(3+\beta)\frac{\delta r}{r}.$$

Solution.
$$\frac{\delta q}{q} = \frac{\delta \rho}{\rho} + \beta \frac{\delta T}{T}.$$

Using $\rho = 3m/4\pi r^3$, we have

$$\frac{\delta \rho}{\rho} = -3\frac{\delta r}{r}.$$

Assuming that the core is described by an ideal gas, we can use Eq. (9.20) with $a = 1$ and $b = 1$, so

$$\left(\frac{4}{3} - a\right)\frac{\delta \rho}{\rho} = \frac{1}{3}\frac{\delta \rho}{\rho} = -\frac{\delta r}{r} = \frac{\delta T}{T}.$$

Combining these two results gives

$$\frac{\delta q}{q} = -3\frac{\delta r}{r} - \beta \frac{\delta r}{r} = -(3+\beta)\frac{\delta r}{r}.$$

11.2 A star starts out with a composition of $X = 0.7$, $Y = 0.26$, and $Z = 0.04$; the metals have a solar distribution. Assume all the hydrogen is burned in the core. What is the Chandrasekhar–Schoenberg limit for this star?

Solution. For the envelope, we have

$$\frac{1}{\mu_{\text{env}}} = \frac{1}{\mu_I} + \frac{1}{\mu_e} = X + \frac{Y}{4} + \frac{Z}{\langle \mathscr{A} \rangle} + \frac{1}{2}(1+X).$$

Solar metallicity has $\langle \mathscr{A} \rangle = 15.5$, so $\mu_{\text{env}} = 0.618$. For the isothermal core, we note that Z doesn't change; therefore $Y = 0.96$, and

$$\frac{1}{\mu_{\text{ic}}} = \frac{1}{2} + \frac{Y}{4} + \frac{Z}{\langle \mathscr{A} \rangle} = 0.7426.$$

Therefore $\mu_{\text{ic}} = 1.346$. Finally, the Chandrasekhar–Schoenberg limit is

$$q = 0.37 \left(\frac{\mu_{\text{env}}}{\mu_{rmic}}\right)^2 = 0.078.$$

11.3 During the transition over the Hertzsprung gap, the radius of a $10\,M_\odot$ star increases from $R_i = 8\,R_\odot$ to $R_f = 250\,R_\odot$. Assume that the core has a mass of $M_{\text{ic}} = 0.1\,M_\odot$ and initial radius of $R_{0\text{ic}} = 0.3\,R_\odot$. Use Eq. (11.35) to show that the radius of the core after the collapse is $0.065\,R_\odot$.

Solution. Starting with

$$\frac{dR}{R^2} = -\left(\frac{M_{\text{ic}}}{M_{\text{env}}}\right)\frac{dR_{\text{ic}}}{R_{\text{ic}}^2},$$

we integrate to find

$$\frac{1}{R_i} - \frac{1}{R_f} = \left(\frac{M_{\text{ic}}}{M_{\text{env}}}\right)\left(\frac{1}{R_{f\text{ic}}} - \frac{1}{R_{0\text{ic}}}\right),$$

where $R_{f\text{ic}}$ is the final core radius. Using $M_{\text{env}} = M - M_{\text{ic}} = 9.9\,M_\odot$, we get

$$R_{f\text{ic}} = 0.065\,M_\odot.$$

Problems of Chapter 12

12.1 Calculate the effective temperature and the core temperature of a $0.7\,M_\odot$ white dwarf if its luminosity is $10^{-3}\,L_\odot$.

Solution. Using the Stefan–Boltzmann law, we have

$$T_{\text{eff}} = \left[\frac{L}{4\pi\sigma R^2}\right]^{1/4} = \left[\frac{10^{-3}\,L_\odot}{4\pi\sigma\,(10^7\,\text{m})^2}\right]^{1/4} = 8{,}570\,\text{K}.$$

Using Eq. (12.19), we have

$$T_c = 10^8 \,\text{K} \left[\frac{1}{1.1} \frac{M_\odot}{M} \frac{L}{L_\odot} \right]^{2/7} = 1.5 \times 10^7 \,\text{K}.$$

12.2 The minimum rotational period of an object held together by gravity is found by equating the tangential velocity at the equator to the orbital velocity of a test particle at the surface. Assume a neutron star remains spherical even at these high rotation rates and determine the minimum orbital period.

Solution. Using Kepler's law and setting $\omega = 2\pi/P$, $a = R = 13.5$ km, and $M = 1.4 M_\odot$, we have

$$P = \sqrt{\frac{4\pi^2 R^3}{GM}} = 7.2 \times 10^{-4}\,\text{s} = 0.72\,\text{ms}.$$

12.3 Consider a pulsar with a mass of $1.4 M_\odot$ and a magnetic field of $B = 10^8$ T at an angle of $\alpha = 30°$.

(a) Assume that the pulsar has constant density and calculate its moment of inertia.

Solution. For a constant density sphere, $I = \frac{2}{5} MR^2$, so

$$I = \frac{2}{5} \left(2.8 \times 10^{30}\,\text{kg}\right) \left(13.5 \times 10^3\,\text{m}\right)^3 = 2.04 \times 10^{38}\,\text{kg m}^2.$$

(b) Define the characteristic lifetime of a pulsar to be $\tau = P/\dot{P}$ and compute the characteristic lifetimes for $P = 100$ ms, $P = 10$ ms, and for the minimum period.

Solution.

$$\tau = \frac{P}{\dot{P}} = \frac{3c^3 \mu_0 I}{8\pi^3 B^2 R^6 \sin^2 \alpha} P^2 = 5.53 \times 10^{15} \, P^2.$$

Therefore,

$$P = 100\,\text{ms} \Longrightarrow \tau = 5.53 \times 10^{13}\,\text{s} = 1.75\,\text{Myr},$$
$$P = 10\,\text{ms} \Longrightarrow \tau = 5.53 \times 10^{13}\,\text{s} = 17{,}500\,\text{year},$$
$$P = 0.72\,\text{ms} \Longrightarrow \tau = 2.87 \times 10^9\,\text{s} = 91\,\text{year}.$$

12.4 Determine the maximum value of the angular momentum that a black hole may have and still have an event horizon.

Solution. The event horizon for a spinning black hole is

$$R_h = \frac{R_s}{2} \left(1 + \sqrt{1 - \frac{4\alpha^2}{R_s^2}} \right).$$

This equation has no solution if $1 - 4\alpha^2/R_s^2 < 0$, so the maximum value of angular momentum occurs at $\alpha = R_s/2$. Using $\alpha = J/Mc$ and $R_s = 2GM/c^2$, we find

$$J = \frac{R_s}{2} Mc = \frac{GM^2}{c}.$$

Problems of Chapter 13

13.1 Assuming no mass loss from the stellar winds, use conservation of angular momentum to compute the semimajor axis of the tidally circularized orbit for step (I) to (II) in the microquasar example. How does it compare to the value of $3,501\,R_\odot$ obtained from the results of the stellar evolution code?

Solution. Starting with the initial semimajor axis of $5,330\,R_\odot$ and eccentricity of 0.6, the tidally circularized orbit will have

$$a' = 5,330\left(1 - (0.36)^2\right) = 3,411\,R_\odot.$$

This is only slightly smaller than the numerically evaluated number.

13.2 Determine the combination of the efficiency parameter and the structure parameter, $\alpha\lambda$ used for the common envelope phase from step (II) to (III), using the values from the microquasar example.

Solution. The combination $\alpha\lambda$ is found from

$$\alpha\lambda = \frac{2m_1 m_1^e a_f a_i}{R_L \left[m_1^c m_2 a_i - m_1 m_2 a_f\right]},$$

where

$$R_L = \frac{0.49 q^{2/3} a_i}{0.69 q^{2/3} + \ln\left(1 + q^{1/3}\right)},$$

with $q = m_1/m_2$. Using $m_1 = 25.5\,M_\odot$, $m_1^c = 15.9\,M_\odot$, $m_2 = 1.22\,M_\odot$, $a_f = 88.2\,R_\odot$, and $a_i = 3,501\,R_\odot$, we find

$$\alpha\lambda = 1.17.$$

13.3 Assume an initial period of 26.1 days for a binary containing a black hole with mass $m_1 = 13.04\,M_\odot$ and a red giant star with mass $m_2 = 1.22\,M_\odot$. If the binary is undergoing stable, conservative mass transfer from the red giant to the black hole with a mass transfer rate of $\dot{m}_2 = 1.0 \times 10^{-8}\,M_\odot/\text{year}$, what are the masses and orbital period after 10.5 Myr?

Solution. Starting with
$$\dot{a} = a\frac{2(m_1 - m_2)}{m_1 m_2}\dot{m}$$
with a constant $\dot{m} = 1.0 \times 10^{-8}\,M_\odot$/year, then $m_1(t) = m_1 + \dot{m}t$ and $m_2(t) = m_2 - \dot{m}t$ with $m_1 = 13.04\,M_\odot$ and $m_2 = 1.22\,M_\odot$, we have
$$\frac{\dot{a}}{a} = \frac{2\dot{m}(\Delta m + 2\dot{m}t)}{m_1 m_2 - \Delta m \dot{m} t - \dot{m}^2 t^2},$$
where $\Delta m = m_1 - m_2$. We integrate from the initial separation a_0 at $t = 0$ to the final separation a at $t = T = 10.5\,\text{Myr}$:
$$\int_{a_0}^{a} \frac{da}{a} = \int_0^T \frac{2\dot{m}(\Delta m + 2\dot{m}t)}{m_1 m_2 - \Delta m \dot{m} t - \dot{m}^2 t^2} dt,$$
so
$$\ln(a)\Big|_{a_0}^a = -2\ln(\dot{m}t(\Delta m + \dot{m}t) - m_1 m_2)\Big|_0^T.$$
This evaluates to
$$a = a_0 \left[\frac{m_1 m_2}{m_1 m_2 - \dot{m}T(\Delta m + \dot{m}T)}\right]^2.$$
Noting that
$$a = \left[\frac{G(m_1 + m_2)}{4\pi^2} P^2\right]^{1/3},$$
we have
$$P = P_0 \left[\frac{m_1 m_2}{m_1 m_2 - \dot{m}T(\Delta m + \dot{m}T)}\right]^3.$$
Plugging in all the appropriate numbers gives $P = 33.4\,\text{d}$ and $\dot{m}T = 0.105\,M_\odot$, so $m_1(T) = 13.145\,M_\odot$ and $m_2(T) = 1.115\,M_\odot$.

Problems of Chapter 14

14.1 Evaluate the gradients in Eq. (14.14) to show that
$$2\sum_i \mathbf{r}_i \cdot \nabla_i \Omega(\mathbf{r}_i) = -2\Omega.$$

Solution. We need to evaluate
$$\nabla \frac{1}{|\mathbf{r} - \mathbf{a}|} = \sum_{i=1}^3 \hat{\mathbf{e}}_i \frac{\partial}{\partial r_i} \frac{1}{\sqrt{(r_1 - a_1)^2 + (r_2 - a_2)^2 + (r_3 - a_3)^2}}$$

where $\hat{\mathbf{e}}_i$ is the unit vector pointing in the ith direction. The derivatives give

$$\nabla \frac{1}{|\mathbf{r}-\mathbf{a}|} = -\frac{(\mathbf{r}-\mathbf{a})}{|\mathbf{r}-\mathbf{a}|^3}.$$

Therefore

$$2\sum_i \mathbf{r}_i \cdot \nabla_i \Omega(\mathbf{r}_i) = 2\sum_{i \neq j} \left[-\mathbf{r}_i \cdot \frac{Gm_i m_j (\mathbf{r}_i - \mathbf{r}_j)}{|\mathbf{r}_i - \mathbf{r}_j|^3} - -\mathbf{r}_j \cdot \frac{Gm_i m_j (\mathbf{r}_j - \mathbf{r}_i)}{|\mathbf{r}_i - \mathbf{r}_j|^3} \right]$$

$$= -2\sum_{i \neq j} \left[(\mathbf{r}_i - \mathbf{r}_j) \cdot \frac{Gm_i m_j (\mathbf{r}_i - \mathbf{r}_j)}{|\mathbf{r}_i - \mathbf{r}_j|^3} \right]$$

$$= -2\sum_{i \neq j} \frac{Gm_i m_j}{|\mathbf{r}_i - \mathbf{r}_j|} = -2\Omega.$$

14.2 Using Eq. (14.20), and an estimate of the density, derive Eq. (14.22).

Solution.

$$t_{\text{rlx}} = \frac{v^3}{8\pi G^2 m_t^2 n \ln(\gamma N)}.$$

Use $m_t n \simeq \langle m \rangle n = \rho$ and $v = \sqrt{GN\langle m \rangle / r}$, so

$$t_{\text{rlx}} = \frac{G^{3/2} N^{3/2} \langle m \rangle^{3/2} r^{-3/2}}{8\pi G^2 \langle m \rangle \rho \ln(\gamma N)}$$

$$= \frac{[N\langle m \rangle / r]^{3/2}}{8\pi \sqrt{G} \langle m \rangle \rho \ln(\gamma N)}$$

$$= \frac{N \left[N\langle m \rangle / r^3 \right]^{1/2}}{8\pi \sqrt{G} \rho \ln(\gamma N)}$$

$$= \frac{N\sqrt{\rho}}{8\pi \sqrt{G} \rho \ln(\gamma N)} = \frac{N}{8\pi \sqrt{G\rho} \ln(\gamma N)}.$$

14.3 Using the density of the Plummer model, compute the core radius r_c using the theoretical definition.

Solution. The Plummer density is

$$\rho(r) = \frac{3Ma^2}{4\pi \sqrt{(r^2 + a^2)^5}},$$

Solutions

so the central density is $\rho(0) = 3M/4\pi a^3$. At the core radius,

$$\rho(r_c) = \frac{\rho(0)}{3} = \frac{M}{4\pi a^3} = \frac{3Ma^2}{4\pi\sqrt{(r_c^2+a^2)^5}}.$$

Solving for r_c gives

$$r_c = a\sqrt{3^{2/5}-1} = 0.743a.$$

14.4 Use the density of the Plummer model.

(a) Compute the surface density of stars using

$$\Sigma(R)R\,dR\,d\phi = \int_{-\infty}^{+\infty} \rho(\sqrt{R^2+z^2})R\,dR\,d\phi\,dz,$$

where the integration is over z. R and ϕ are cylindrical coordinates.

Solution.

$$\Sigma(R)R\,dR\,d\phi = \frac{3Ma^2}{2\pi}R\,dR\,d\phi\int_0^\infty (R^2+a^2+z^2)^{-5/2}\,dz.$$

The integral evaluates to

$$\int (R^2+a^2+z^2)^{-5/2}\,dz = \frac{1}{(R^2+a^2)^2}\left[\frac{z}{\sqrt{R^2+a^2+z^2}} - \frac{1}{3}\frac{z^3}{\sqrt{(R^2+a^2+z^2)^3}}\right].$$

Taking the appropriate limits, we have

$$\Sigma(R) = \frac{Ma^2}{(R^2+z^2)^2}.$$

(b) Compute the core radius r_c using the observational definition.

Solution. Using the surface density computed in (a), the central surface density is

$$\Sigma(0) = \frac{M}{\pi a^2}.$$

From the observational definition of the core radius,

$$\Sigma(R_c) = \frac{1}{2}\Sigma(0) = \frac{M}{2\pi a^2} = \frac{Ma^2}{\pi(R_c^2+a^2)^2}.$$

Solving for R_c gives

$$R_c = a\sqrt{\sqrt{2}-1} = 0.644a.$$

Problems of Chapter 15

15.1 Calculate the average and median values of the eccentricity for a population of binaries with a thermal distribution of eccentricities.

Solution. The mean is found from the integral:

$$\langle e \rangle = \int_0^1 e 2e\,de = 2\int_0^1 e^2\,de = \frac{2}{3} \simeq 0.67.$$

The median (e_m) is found by solving the following integral equation:

$$\frac{1}{2} = \int_0^{e_m} 2e\,de = e_m^2 \implies e_m = \frac{\sqrt{2}}{2} \simeq 0.71.$$

15.2 Assume a globular cluster consists of stars with mass $m = 0.7\,M_\odot$ and has a dispersion velocity of $\langle v \rangle = 20\,\text{km/s}$. Determine the maximum orbital period of a "hard" binary.

Solution. From the virial theorem, the relative binding energy of a binary is

$$x = -\frac{1}{2}\mu v_{\text{orb}}^2 + \frac{GM\mu}{r} = \frac{Gm^2}{2a},$$

where a is the semi-major axis of the orbit, v_{orb} is the orbital velocity, and $M\mu = m^2$ for equal mass systems. The dynamical temperature of this cluster is

$$kT = \frac{1}{3}m\langle v \rangle^2.$$

Therefore the critical binding energy for a hard binary is found from $x = kT$, and so:

$$\frac{Gm^2}{2a} = \frac{1}{3}m\langle v \rangle^2.$$

Solving for a and using Kepler's third law, we have:

$$a = \frac{3Gm}{2\langle v \rangle^2} = \left(\frac{2GmP^2}{4\pi^2}\right)^{1/3}.$$

Solving for P gives

$$P = \frac{3\sqrt{3}\pi Gm}{2\langle v\rangle^3} = 9.53 \times 10^7\,\mathrm{s} = 3.02\,\mathrm{year}.$$

15.3 Solve Eqs. (15.20) and (15.21) to obtain expressions for $(\Delta t)^2 \ddot{\mathbf{a}}(t)$ and $(\Delta t)^3 \dddot{\mathbf{a}}(t)$ in terms of $\mathbf{a}(t)$, $\mathbf{a}(t+\Delta t)$, $\dot{\mathbf{a}}(t)$, and $\dot{\mathbf{a}}(t+\Delta t)$.

Solution. Start with

$$\mathbf{a}(t+\Delta t) = \mathbf{a}(t) + \Delta t \dot{\mathbf{a}}(t) + \frac{1}{2}(\Delta t)^2 \ddot{\mathbf{a}}(t) + \frac{1}{6}(\Delta t)^3 \dddot{\mathbf{a}}(t),$$

$$\dot{\mathbf{a}}(t+\Delta t) = \dot{\mathbf{a}}(t) + \Delta t \ddot{\mathbf{a}}(t) + \frac{1}{2}(\Delta t)^2 \dddot{\mathbf{a}}(t).$$

Multiply the second equation by $(\Delta t)/3$ to get

$$\frac{1}{3}(\Delta t)\dot{\mathbf{a}}(t+\Delta t) = \frac{1}{3}(\Delta t)\dot{\mathbf{a}}(t) + \frac{1}{3}(\Delta t)^2 \ddot{\mathbf{a}}(t) + \frac{1}{6}(\Delta t)^3 \dddot{\mathbf{a}}(t)$$

and subtract this from the first equation to find

$$\mathbf{a}(t+\Delta t) - \frac{1}{3}(\Delta t)\dot{\mathbf{a}}(t+\Delta t) = \mathbf{a}(t) + \frac{2}{3}(\Delta t)\dot{\mathbf{a}}(t) + \frac{1}{6}(\Delta t)^2 \ddot{\mathbf{a}}(t).$$

Solving for $(\Delta t)^2 \ddot{\mathbf{a}}(t)$ gives

$$(\Delta t)^2 \ddot{\mathbf{a}}(t) = 6\left(\mathbf{a}(t+\Delta t) - \mathbf{a}(t)\right) - 2(\Delta t)\left(\dot{\mathbf{a}}(t+\Delta t) + 2\dot{\mathbf{a}}(t)\right).$$

Substitute this back int either of the first two equations to find

$$(\Delta t)^3 \dddot{\mathbf{a}}(t) = -12\left(\mathbf{a}(t+\Delta t) - \mathbf{a}(t)\right) + 6\Delta t\left(\dot{\mathbf{a}}(t+\Delta t) + \dot{\mathbf{a}}(t)\right).$$

Index

Symbols
α-prescription, 189
γ-prescription, 189
H^- opacity, 142

A
Absolute magnitude, 11
Absorption line, 38
Absorption lines, 30
Adiabatic exponent, 73
Adiabatic temperature gradient, 131
Aldebaran, 8
Angle of inclination, 21, 34
Apastron, 17
Apparent magnitude, 11
Ascending node, 21
Astrometric binary, 13
Astronomical unit, 29
Asymptotic giant branch, 156, 158
Atomic mass unit, 52
Auxiliary circle, 19
Azimuthal angle, 6

B
Barycenter, 17
Beta decay, 170
Binary burning, 220
Binary stars, 13
Blackbody specific intensity, 78
Blackbody spectrum, 30, 33
Boltzmann equation, 38
Bound-free opacity, 85
Brehmstrahlung, 81

C
Carbon burning, 100
Celestial coordinates, 3, 6
Celestial equator, 6
Chandrasekhar mass, 112
Chandrasekhar-Schoenberg limit, 152
CNO cycle, 97
CNO II cycle, 98
CNO-I cycle, 98
Co-latitude, 6
Collisional Boltzmann equation, 207
Collisionless Boltzmann equation, 204
Color-Magnitude diagram, 34
Common envelope, 189
Composition equation, 54
Convection, 126
Convective flux, 129
Core collapse, 211
Core radius, 197
Coulomb logarithm, 200
Crossing time, 197
Crystallization, 168
Curvature radiation, 173

D
Declination, 6
Degeneracy condition, 69
Degeneracy equation of state, 70
Degeneracy pressure, 68
Degenerate state, 40
Density of states, 32
Dissociation, 139
Distance modulus, 12
Distribution function, 203

Doppler shift, 13, 25
Dredge up, 155
Dynamical instability, 124
Dynamical temperature, 210
Dynamical timescale, 59

E
Eccentric anomaly, 15, 19
Eccentricity, 17, 21, 35
Eclipsing binary, 13, 34
Ecliptic, 7
Eddington approximation atmosphere, 88
Eddington luminosity, 117
Eddington standard model, 115
Effective temperature, 34
Electron pressure, 66
Electron scattering opacity, 85
Emission coefficient, 80
Energy equation, 49
Energy flux, 79
Epoch, 7
Equation of state, 63
Equatorial coordinates, 6
Ergosphere, 177
Evaporation time, 201
Evolution equations, 54, 63

F
Fermi energy, 69
Fermi momentum, 72
Fragmentation, 136
Free-fall timescale, 134
Free-free opacity, 85

G
Gas constant, 66
Gas pressure, 65
Gravitational potential energy, 55
Gravothermal catastrophe, 211
Grey atmosphere, 86

H
Half-light radius, 198
Half-mass radius, 198
Hayashi track, 144, 155
Heggie-Hills law, 217
Helium flash, 157
Henyey track, 145
Hertzsprung gap, 155
Hertzsprung-Russell diagram, 34

Hipparchus, 11
Horizontal branch, 155
Hydrodynamic equation, 50
Hydrostatic equilibrium, 119

I
Ideal gas law, 51, 63
Intensity, 30
Internal energy, 71
Invariance, 175
Ion pressure, 65
Ionization state, 41
Iron core, 160
Isothermal collapse, 136
Isothermal core, 149

J
Jeans criteria, 133
Jeans mass, 135

K
Kepler's 2nd Law, 21
Kepler's 3rd Law, 17
Kepler's equation, 18
Kerr metric, 176
Kramers opacity, 85

L
Lagrange points, 183
Lane-Emden equation, 109
Light curve, 34
Light cylinder, 173
Limb, 36
Limb darkening, 36, 89
Local thermodynamic equilibrium, 38, 48, 88
Longitude of the ascending node, 21
Longitude of the periastron, 21, 35
Luminosity, 11, 30

M
Magnetosphere, 173
Magnitude scale, 11
Mass excess, 94
Mass fraction, 51, 65
Mass function, 27
Mass segregation, 208
Mass transfer – case A, 186
Mass transfer – case B, 186
Mass transfer – case C, 186

Mass-Luminosity relation, 30
Maxwell-Boltzmann distribution, 38, 67
Mean atomic mass, 65
Mean intensity, 77
Mean molecular weight, 65
Meridian, 4
Metallicity, 51, 65
Metric, 175
Micro quasar, 191
Mixing length, 129

N
Neutron capture, 101
Neutron drip, 170
Neutron stars, 112
Neutronization, 170
Newton-Raphson, 21
Normalized potential, 182
North celestial pole, 6
Nuclear fusion, 51, 91
Nuclear timescale, 60

O
Oblateness, 214
Opacity, 80
Opacity Project, 84
Optical depth, 81
Orbital elements, 21, 22, 35
Orbital velocity, 35
Oxygen burning, 100

P
P-P I chain, 95
P-P II chain, 96
P-P III chain, 96
Parallax, 29
Parsec, 12, 29
Periastron, 17
Photodisintegration, 101
Photons, 30
Photosphere, 34, 38
Planck distribution, 33, 67
Plane-parallel atmosphere, 85
Plummer model, 206
Plummer radius, 206
Polytropes, 108
Polytropic equation of state, 108
Polytropic index, 108
Precession of the equinox, 5, 7
Pressure integral, 67
Proper motion, 8

Proton-proton chain, 95
Protostar, 138
Pseudopotential, 182

R
R-process, 102
Radial motion, 8
Radiant flux, 11
Radiation constant, 67
Radiation pressure, 65, 79
Radiative heat flux, 85
Radius-mass exponent, 185
Reaction rate, 94
Relaxation time, 198
Resonance peak, 94
Right ascension, 6
Roche lobe, 183
Roche lobe overflow, 184
Roche lobe radius, 184
Rosseland mean opacity, 83

S
S-process, 102
Saha equation, 38, 41
Schwarzschild metric, 175
Schwarzschild radius, 175
Secular thermal instability, 122
Semi-amplitude of the velocity, 25
Semi-major axis, 17, 21
Shell flash, 159
Shockwave, 138
Sidereal day, 4
Sidereal month, 4
Sidereal year, 5
Solar day, 4
Source function, 80
South celestial pole, 6
Specific angular momentum, 19
Specific intensity, 77
Spectral class, 38
Spectroscopic binary, 13, 24
Spherical law of cosines, 9
Spherical law of sines, 9
Spitzer instability, 208
Stefan constant, 34
Stefan-Boltzmann constant, 67
Stefan-Boltzmann laws, 34
Stellar atmosphere, 85
Structure equations, 107
Supernova, 162
Synchrotron radiation, 173
Synodic month, 4

T
Thermal eccentricity, 215
Thermal timescale, 59
Thin shell instability, 123
Tidal radius, 198
Time of periastron, 21
Transfer equation, 80
Triple alpha process, 99
True anomaly, 19

U
URCA process, 172

V
Vernal equinox, 5
Violent relaxation, 208
Virial theorem, 55
Visual binary, 13, 28

W
White dwarfs, 112

Z
ZAMS, 145, 147

MIX
Papier aus verantwortungsvollen Quellen
Paper from responsible sources
FSC® C105338

If you have any concerns about our products,
you can contact us on
ProductSafety@springernature.com

In case Publisher is established outside the EU,
the EU authorized representative is:
**Springer Nature Customer Service Center GmbH
Europaplatz 3, 69115 Heidelberg, Germany**

Printed by Libri Plureos GmbH
in Hamburg, Germany